The Life, Extinction, and Rebreeding of Quagga Zebras

Significance for Conservation

Quaggas were beautiful pony-sized zebras in southern Africa that had fewer stripes on their bodies and a browner body coloration than other zebras. Indigenous people hunted quaggas, portrayed them in rock art, and told stories about them. Settlers used quaggas to pull wagons and to protect livestock against predators. Taken to Europe, they were admired, exhibited, harnessed to carriages, illustrated by famous artists, and written about by scientists. Excessive hunting led to quaggas' extinction in the 1880s but DNA from museum specimens showed rebreeding was feasible and now zebras resembling quaggas live in their former habitats. This rebreeding is compared with other de-extinction and rewilding ventures and its appropriateness is discussed against the backdrop of conservation challenges – including those facing other zebras. In an Anthropocene of species extinction, climate change, and habitat loss, which organisms and habitats should be saved, and should attempts be made to restore extinct animals?

PETER HEYWOOD is Professor of Biology in the Department of Molecular Biology, Cell Biology and Biochemistry at Brown University, USA, where he has taught since 1974. He is a fellow of the Linnean Society and a fellow of the Royal Society of Biology. Most of his scholarship has focused on the cell biology of algae, protists, and animals, but he has also published on development of the inner ear in mammals, agricultural biotechnology, pedagogy, and biography. His interest in quaggas started in 2006, and he has written on the history of quagga zebras, their representations in biology, art, and literature, and their rebreeding.

ECOLOGY, BIODIVERSITY AND CONSERVATION

The world's biological diversity faces unprecedented threats. The urgent challenge facing the concerned biologist is to understand ecological processes well enough to maintain their functioning in the face of the pressures resulting from human population growth. Those concerned with the conservation of biodiversity and with restoration also need to be acquainted with the political, social, historical, economic and legal frameworks within which ecological and conservation practice must be developed. The new Ecology, Biodiversity and Conservation series will present balanced, comprehensive, up-to-date and critical reviews of selected topics within the sciences of ecology and conservation biology, both botanical and zoological, and both 'pure' and 'applied'. It is aimed at advanced final-year undergraduates, graduate students, researchers and university teachers, as well as ecologists and conservationists in industry, government and the voluntary sectors. The series encompasses a wide range of approaches and scales (spatial, temporal and taxonomic), including quantitative, theoretical, population, community, ecosystem, landscape, historical, experimental, behavioural and evolutionary studies. The emphasis is on science related to the real world of plants and animals rather than on purely theoretical abstractions and mathematical models. Books in this series will, wherever possible, consider issues from a broad perspective. Some books will challenge existing paradigms and present new ecological concepts, empirical or theoretical models, and testable hypotheses. Other books will explore new approaches and present syntheses on topics of ecological importance.

Ecology and Control of Introduced Plants
Judith H. Myers and Dawn Bazely

Invertebrate Conservation and Agricultural Ecosystems
T. R. New

Risks and Decisions for Conservation and Environmental Management
Mark Burgman

The Life, Extinction, and Rebreeding of Quagga Zebras

Significance for Conservation

PETER HEYWOOD

Brown University

 CAMBRIDGE
UNIVERSITY PRESS

CAMBRIDGE
UNIVERSITY PRESS

University Printing House, Cambridge CB2 8BS, United Kingdom

One Liberty Plaza, 20th Floor, New York, NY 10006, USA

477 Williamstown Road, Port Melbourne, VIC 3207, Australia

314–321, 3rd Floor, Plot 3, Splendor Forum, Jasola District Centre, New Delhi – 110025, India

103 Penang Road, #05–06/07, Visioncrest Commercial, Singapore 238467

Cambridge University Press is part of the University of Cambridge.

It furthers the University's mission by disseminating knowledge in the pursuit of
education, learning, and research at the highest international levels of excellence.

www.cambridge.org
Information on this title: www.cambridge.org/9781108831604
DOI: 10.1017/9781108917735

First published 2022

Printed in the United Kingdom by TJ Books Limited, Padstow Cornwall

A catalogue record for this publication is available from the British Library.

Library of Congress Cataloging-in-Publication Data
Names: Heywood, Peter (Biologist), author.
Title: The life, extinction, and rebreeding of quagga zebras : significance for conservation / Peter
  Heywood, Brown University, Rhode Island.
Description: Cambridge, United Kingdom ; New York, NY : Cambridge University Press, 2022. |
  Series: Ecology, biodiversity and conservation | Includes bibliographical references and index
Identifiers: LCCN 2021060144 (print) | LCCN 2021060145 (ebook) | ISBN 9781108831604
  (hardback) | ISBN 9781108926911 (paperback) | ISBN 9781108917735 (epub)
Subjects: LCSH: Quagga–Africa, Southern. | Quagga–Conservation. | Quagga–Ecology. |
  Quagga–Breeding. | BISAC: NATURE / Ecology
Classification: LCC QL737.U62 H49 2022 (print) | LCC QL737.U62 (ebook) |
  DDC 599.665/7168–dc23/eng/20211216
LC record available at https://lccn.loc.gov/2021060144
LC ebook record available at https://lccn.loc.gov/2021060145

ISBN 978-1-108-83160-4 Hardback
ISBN 978-1-108-92691-1 Paperback

For Nancy, Ela, Wesley, Margaret, Ray, Hazel, Amos, Simon, and Marie

Contents

Color plates can be found between pages 120 and 121

Acknowledgments

My interest in quaggas began about fifteen years ago and there are many debts to be gratefully acknowledged. During this time (and for decades before) I taught at Brown University, which has provided me with sabbaticals and the facilities for scholarship. I am grateful to the administrators, colleagues, staff, and students who have aided me in many ways. The Brown University librarians have been unfailingly helpful, in particular Karen Bouchard, Ann Dodge, Frank Kellerman, Gayle Lynch, William Munroe, Erika Sevetson, and Kimberly Silva. The Pembroke Center Research Seminar, "Visions of Nature," provided me with a broader context for considering wildlife; I am grateful to the other participants and the leader, Leslie Bostrom.

My visits to South Africa were made fruitful by the staff of the Iziko Museum, particularly Denise Hamerton and Vicky MacCrae; by Eric Harley, March Turnbull, and Bernard Wooding of the Quagga Project; and by Andrew Bank and Anja Macher, who helped me greatly. Special thanks are due to March Turnbull, who has provided valuable assistance and information over the years. I am indebted to Reinhold Rau – whom, unfortunately, I never had the pleasure of meeting – for publishing information about quaggas and for initiating the Quagga Project; without his work this book would have been incomplete.

I thank those people who answered my questions: Alan Barnard, Denise Hamerton, Graham Kerley, Sarah King, Hannes Lerp, Patricia Moehlman, David Morris, Carsten Renker, Oliver Ryan, Reinier Spreen, Keenan Stears, and Andrew Taylor. I owe a debt of gratitude to those who read and commented on sections of my manuscript: Saeideh Esmaeili, Megan Gura, Jacob Hennig, Patrick Malone, Peter Novellie, Dov Sax, March Turnbull, and Bernard Wooding; needless to say, the mistakes in this book are mine alone.

For editorial help, I am indebted to Susan Dearing, Priscilla Hall, Nancy Jacobs, Cathy Munro, Aleksandra Serocka, Jenny van der Meijden, Sindhuja Sethuraman, Indra Priyadarshini and Liz Steel. I am grateful to Judi Gibbs for preparing the index. I thank the series editor, Michael Usher, for seeking me out and guiding me through the process of preparing a book proposal.

The figures are from a variety of sources. I am grateful to Bruce Boucek and Camille Tulloss for preparing illustrations, to the librarians who photographed images that are no longer copyrighted, and to British Museum Images, the Rijksmuseum, the Royal College of Surgeons of England, the Teylers Museum, the National Archives and Record Services of South Africa, the Western Cape Provincial Parliament, the Quagga Project and the Zoological Society of London for the use of their images. I thank the following people for the use of their images: T. Bobosh, Yathin S. Krishnappa, Brenda Larison et al. and Bernard Wooding.

As this may be the only book that I'll ever write, I use this opportunity to record my gratitude to the broader circle of people who made my research and teaching possible. My thanks go to my parents and grand-parents and the many people in Britain who supported and encouraged me, and to my teachers at Crumpsall Lane Primary School, Crumpsall Methodist Church, and North Manchester Grammar School who pre-pared me for a lifetime of learning. I am grateful to the faculty and staff of Queen Mary College, University of London, where I received my undergraduate and graduate degrees, and to the many people who have helped me subsequently.

For more than forty years, I have met regularly with Bill, Don, Harold, Howard, Mark, Mark, Patrick, and Rich; their friendship has been very important, and I record their names with affection. People that I never knew personally but who have improved my life in different ways include: Vasily Arkhipov, Hilda Porter, Richard B. Solomon and his wife, and Colin Sullivan.

Our children, Ela and Wesley, have had to listen to too many conver-sations about quaggas; I am grateful to them for their patience and for the happiness they have brought me. I dedicate this book to them, to my wife, Nancy, and to my sister, Margaret, and her family. Being in Africa with Nancy sparked new interests and she has been at my side through the research and writing of this book – alerting me to sources, comment-ing on chapters, being my computer consultant, and encouraging me. Life would have been very different without Nancy; I thank her with all my heart, and I look forward to our work, travel, and time together in the years to come.

Introduction

The name of this species is derived from its voice, which is a kind of cry somewhat resembling the sounds qua-cha! It is unquestionably best calculated for domestication, both as regards strength and docility ... we have ourselves been drawn by one in a gig, the animal showing as much temper and delicacy of mouth as any domestic horse ... It is this species that is reputed to be the boldest of all Equine animals, attacking hyaena and wild dog without hesitation, and therefore not unfrequently domesticated by the Dutch boors for the purpose of protecting their horses at night while both are turned out to grass.

C. Hamilton Smith[1]

Such is a thumbnail sketch of an animal the size of a pony that became extinct in the late nineteenth century but whose use in traction and livestock protection might have led to domestication. Quaggas were distinctive in their appearance: they had fewer stripes than other zebras and these were dark reddish-brown or black and a lighter reddish-brown.[2] Stripes were confined to their faces, necks, manes and forequarters, before becoming irregular and then disappearing altogether at the hindquarters, that were a reddish-brown color. Their unstriped legs were white, as were their underbodies, except for a ventral stripe along their length (Figure 0.1). They seemed to have evolved this distinctive coat coloration from living in an isolated habitat that was colder and drier than the environments of most zebras.

Quaggas lived in a limited area of southern Africa that extended from east of present-day Cape Town across the arid plateau of the Karoo to the grasslands of the high veld – areas that became part of the Cape Colony and the Orange Free State (Figure 0.2). Their preferred food was grass, but they could eat other vegetation. They lived in breeding groups comprising a stallion, one or more mares, and their foals.[3] Several groups might join to form a herd and move together looking for water and

Figure 0.1. The quagga in the Zoological Society's Gardens, Regent's Park. From Weir, The Illustrated London News, 1858

forage. Sometimes, large numbers congregated: one nineteenth-century account describes "bands of many hundreds" migrating in search of new grazing, and another describes how multitudes of quaggas and other animals picked clean the vegetation in the path of their migration.[4]

Quaggas were not the only zebras in southern Africa: to the north and east were animals that resembled quaggas, but had more stripes and were less brown, and scattered through the west and south-west were mountain zebras, abundantly marked with black and white stripes. Quaggas shared grazing with several species of antelopes, ostriches, and wildebeests (gnus), but it was with the latter two that they formed close bonds that seem to reflect feeding preferences. All these animals faced predation by lions, leopards, hyenas, cheetahs, wild dogs, and humans. The indigenous Khoe-San, who lived alongside quaggas for millennia, immortalized them in paintings and engravings, while also hunting them for meat, hides, and bones.[6] Lacking firearms and horses, they relied on skill and ingenuity to make and use bows, arrows, spears, poison, and hunting traps. The |Xam (hunter-gatherers of the Karoo) told stories about

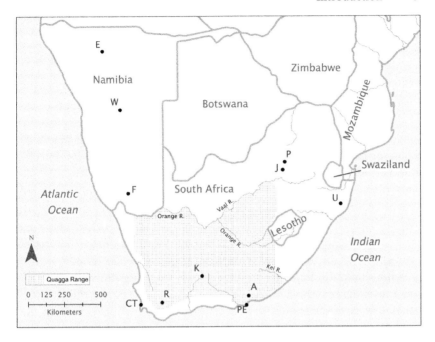

Figure 0.2. Southern Africa, showing locations mentioned in this book. Quaggas lived in areas that are now part of the Western Cape, Eastern Cape, Northern Cape, and the Free State provinces; stippling shows their presumed range in the mid-eighteenth century. The Fish River Canyon, Namibia (F) was a site of unsuccessful searches for quaggas after their extinction. Plains zebras from Umfolozi Game Reserve, South Africa (U) and Etosha National Park, Namibia,(E) were brought to the Vrolijkheid Nature Conservation Station near Robertson (R), the initial location of the Quagga Project, and their offspring now live at several locations, including Addo Elephant National Park (A) and Karoo National Park (K). The map indicates the following cities: Cape Town (CT), Johannesburg (J), Port Elizabeth (PE), Pretoria (P), and Windhoek (W). Swaziland is now Eswatini. This figure by the author was prepared by Bruce Boucek of the Brown University Library (courtesy of Brown University Library); it was originally published in Heywood (2015) and is reprinted courtesy of Kronos[5]

quaggas in which they are not just hunted animals, but also sentient beings living in households with families.[7]

When first encountered, quaggas were wondrously novel to the Dutch at the Cape, who wondered if they might substitute for the domestic equines that they sorely needed. Eventually, they were used to pull wagons and carriages in the Cape Colony and Britain, and some farmers kept quaggas with livestock to protect them against predators by biting and kicking aggressors that came close. Obtained young and properly treated, quaggas could have been widely used and probably would have been more resistant to tropical diseases than horses.

Their potential for protection or draft, however, did not prevent quaggas from being killed in large numbers by farmers and hunters in a region engulfed in the eighteenth and nineteenth centuries by the expanding frontier of the Cape Colony. Some people shot quaggas frivolously for "sport"; others killed them to provide meat for servants and laborers and for their hides, which were used to make rough rawhide shoes, leather thongs for binding things together, and cheap bags. Finally, they became the prey of hunters who exported their hides to provide high-quality leather for overseas boot makers. It was all too much: hunted extensively and excluded from water and grazing lands increasingly used to support farm livestock, quaggas that had existed for many millennia became extinct in the wild in the 1870s, leaving the endling – the last of their kind – to die in the Amsterdam Zoo on August 12, 1883.[8]

Some organisms become extinct even before being known to science and the wider world. This was not the case for quaggas: they shaped Charles Darwin's ideas about mechanisms of inheritance, justified the plot elements of August Strindberg's play *The Father*, and featured in the stories of the |Xam. Portrayed by unknown rock artists and famous European artists such as Jacques-Laurent Agasse and Nicolas Maréchal, we can envision them today.

Cape mountain zebras, *Equus zebra zebra*, almost suffered the fate of quaggas. People continued to hunt them into the twentieth century when fewer than ninety survived.[9] In 1937, authorities designated a farm containing Cape mountain zebras near Cradock in the Eastern Cape as the Mountain Zebra National Park.[10] In 1950, conservationists moved five stallions and six mares from other locations to the park and a few more animals were added in 1964. Since then, their numbers have steadily increased into the thousands.[11]

Bonteboks, *Damaliscus pygargus pygargus*, have a similar success story. They, too, like quaggas and Cape mountain zebras, were endemic to South Africa. By 1931, only twenty-two animals remained on a single farm, but conservation has increased their numbers into the thousands, many of which live in the Bontebok National Park.[12] Common to both these accounts is that dedicated people saved a subspecies from extinction by providing the animals with safe, suitable habitats. Initially, this conservation involved neither extensive areas nor major expenditures. Most people take the existence of these animals for granted, not realizing how close they came to extinction.

Central to the success of conservation in South Africa has been the preservation of habitats such as Mountain Zebra National Park and Kruger National Park, linked to national parks in Mozambique and Zimbabwe to form the Great Limpopo Transfrontier Park.[13] Some other reserves are also large and provide wildlife corridors that allow migration. Extensive reserves allow sizeable populations of individual species and so reduce the possibility of inbreeding.

To return to quaggas, their story did not end with extinction. Reinhold Rau, a taxidermist at the South African Museum in Cape Town, sent tissue samples from quagga hides to scientists in the United States for analysis in the early 1980s, and these little-known zebras gained the distinction of being the first extinct creatures to have short segments of their DNA sequenced. Quagga DNA sequences proved to be very similar to sequences found in plains zebras, which led to reclassification: quaggas were not a separate species, as some taxonomists had concluded, but were a type of plains zebra that now bore the binomial name, *Equus quagga.*

Rau reasoned that, because quaggas were a subspecies of plains zebras, the genes for their distinctive coat coloration might still exist within populations of plains zebra and be retrievable by selective breeding. He knew that there were populations of plains zebras in both South-West Africa (present-day Namibia, but then under South African administration) and South Africa that resembled quaggas in their reduced striping and coat coloration, and in 1987 he obtained some of these animals as the founder population for the "Quagga Project" – a venture that has received the enthusiastic support of many individuals and organizations.[14] Careful selection of the descendants of these animals by the Quagga Project has yielded zebras with fewer stripes and more brown coloration that now live in the same habitats once occupied by quaggas.

At first sight, this outcome represents an inspiring success story, but it raises the question: to what extent *are* the rebred animals quaggas? Even the brownest of them do not have the chestnut color visible in many paintings, and – importantly – were there genetic, morphological, and behavioral characteristics of quaggas that are not present in Quagga Project zebras?

The opportunity cost of rebreeding is another issue: could the money and resources devoted to the Quagga Project have been used more effectively in conservation projects? South Africa is a country of great biodiversity, but the International Union for Conservation of Nature (IUCN) lists many of its species as endangered. Losing large vertebrate species to hunters may be a thing of the past, but climate change and habitat loss all take their toll as the human population expands and as people convert wildlife habitats into housing, farmers' fields, wineries, mines, etc. Might some funding that went to rebreeding have been used to preserve these habitats? Extinction and rebreeding raise other important questions. How do humans value species and subspecies? What combinations of circumstances cause anthropogenic extinctions? What organisms can be saved, and how can this best be done? Is rebreeding extinct animals a worthwhile endeavor in a world of pressing conservation needs?

Whatever the success of rebreeding, people remember the original quaggas, as they have inspired stories, poems, illustrations, and an animated film. Humans have given their name to other animals and to a range of enterprises. Descriptions from the |Xam, explorers, hunters, scientists, and poets in this book present authentic accounts of these zebras. Their voices will help to tell the story of how quaggas lived, why they were lost, and how people are attempting to rebreed them. The most prominent presence, nevertheless, should be that of a remarkable animal that was an important part of its environment and whose extinction continues to serve as a warning about the dire effects of human greed and folly on the natural world.

1 · *Zebras*

I stopped to examine these zebras with my pocket telescope: they were
the most beautifully marked animals I had ever seen: their clean sleek
limbs glittered in the sun, and the brightness and regularity of their striped
coat, presented a picture of extraordinary beauty ...

William John Burchell[1]

Plains zebras, mountain zebras, and quaggas were important members of
the southern African biota: they interacted with other grazing animals
and, together with them, were prey for predators. Collectively, these
animals influenced the nature of several biomes.

Zebras and quaggas were well known to indigenous people: they
provided them with necessities, such as hides and meat, and featured in
their art and stories. Indigenous people distinguished between quaggas
and mountain zebras. To the |Xam, the first known inhabitants of the
Karoo, they were "‖ʊkhwĩ" and "‖kabba," respectively.[2] People who
spoke the Xhosa language seem to have differentiated between them by
their calls: they called quaggas with their barking neigh "iqwarha," and
knew Cape mountain zebras as "idawuwa."[3] The other indigenous people
of the region, the Khoekhoe, named both mountain zebras and quaggas
"quacha" (sometimes rendered as "qua-cha" or "quakka").[4] "Iqwarha"
and "quacha" − like the Dutch and English names of "kwagga" and
"quagga" − are onomatopoeic, suggesting the animal's barking cry of
"kwa-haa, kwa-haa," which was often repeated in rapid succession.

After the European colonists introduced horses, which in Xhosa were
"ihashe," to the region, the Xhosa incorporated a reference to the equine
family into their name for the barking zebras: "iqwarhashe."[5] By differ-
entiating local animals and connecting them to a global family, those
names − iqwarha and iqwarhashe − represent the project of taxonomy
well, but writ small. Globally, there were several horse-like animals,
some striped, some not, and clear names are necessary to catalog the set.

Equines

Carl Linnaeus (1707–1778) developed the modern system of classification that uses a hierarchy of categories to construct closer and more distant affinities. Species are grouped most closely within a genus; the binomial name of genus and species provides a signifier that is unique to the organism. Because many reports of animals came to Linnaeus from all over the world, he had a bigger challenge than Xhosa speakers did with their handful of local species. Linnaeus had to sort out several types of horse-like creatures. Recognizing the similarity between horses, zebras, and asses (wild donkeys), he placed them in the same genus, which he called *Equus*, using the Latin name for horse. Collectively, these animals are referred to as equines. The genus *Equus* is grouped with extinct equids into the family Equidae.[6]

The ancestors of the Equidae can be traced back over 50 million years to animals less than twenty inches (508 mm) tall and with three toes on their hind feet and four toes on their fore feet. The earliest equids browsed on various plants but, with the creation of vast grasslands, their descendants evolved into larger grass-eating animals. The fossil record shows that there were many such species, most of which are now extinct. Equines, animals clearly belonging to the *Equus* lineage, were present in North America about 4 million years ago and their descendants living there died out about 10,000 years ago. Beginning about 3 million years ago, equines spread into South America and, via the land bridge of Beringia, into Eurasia and Africa where their descendants became horses, asses, and zebras.[7]

Equines use their incisors to crop vegetation and their premolars and molars to grind it; their digestive systems enable them to live by eating large amounts of food having a low nutrient content. Their eyes on the sides of their heads afford vision over a wide arc to detect predators, and their senses of hearing and smell provide additional warning of dangers.[8] Speed to avoid predators comes from highly specialized legs whose digits are reduced to just the middle toe of each foot, which is thickened and greatly elongated; their long legs end in hard, impact-absorbing hooves.[9] Zebras can run at speeds up to 55 kilometers per hour (34 miles per hour) and horses are even faster.[10] As well as providing speed, strong legs can deliver a powerful kick to a predator or competitor.

DNA evidence reveals the evolutionary history of horses, asses, and zebras. The sequence of four nucleotides, chemical groups that are linked end-to-end to form DNA molecules present in nuclei and mitochondria, provides key information. Collectively, the DNA of an organism is called its genome and its study is termed genomics. The technique of DNA sequencing enables scientists to compare the order of nucleotides between the genomes of different organisms: an identical or similar sequence of nucleotides argues that two organisms are closely related, or even belong to the same species. Scientists study genomes from both extant (present-day) equines and from preserved material including ancient horse bones preserved in permafrost for over 500,000 years and quagga hides and bones.[11] Genomes evolve over time and so the degree of difference between nucleotide sequences of species can help determine the time when they diverged from a common ancestor.

The genus Equus diverged into evolutionary lineages leading to asses, zebras, and horses (Figure 1.1). One horse, *Equus ferus caballus*, was

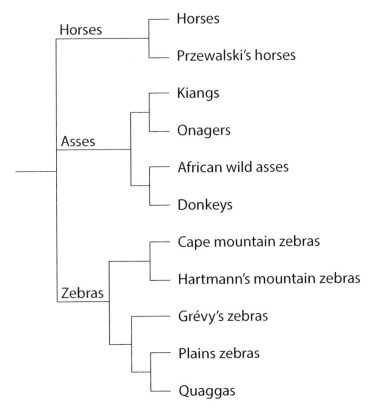

Figure 1.1. Evolutionary relationships between equines showing the divergence of lines leading to horses, asses, and zebras. This figure by the author was prepared by Camille Tulloss[12]

domesticated about 5,000–6,000 years ago and another, Przewalski's horse, *Equus ferus przewalskii*, remains wild and endangered. Asses exist in Asia as *Equus kiang* (kiangs) and *Equus hemionus* (Asian wild asses or onagers) and in Africa as wild asses (*Equus africanus*) whose domestication gave rise to donkeys (*Equus africanus asinus*).

Hippotigris[13]

Linnaeus took the common name *zebra* – derived from a Portuguese word meaning "wild ass" – and used it as the species name for the striped equine that lived in mountainous areas of southern Africa, *Equus zebra*. Boddaert and Gmelin distinguished this species from the partially striped equine of the plains, which they named *Equus quagga* – basing the name on the animal's barking call.[14]

Hippotigris, the name Romans gave to zebras, is used to denote a subgenus of *Equus* containing the three zebra species.[15] It is thought that the ancestors of zebras diverged from the ancestors of asses approximately 1.7 to 2.0 million years ago and that the common ancestor of all Hippotigris species separated between 1.28 to 1.59 million years ago into the evolutionary lines leading to the currently recognized species: mountain zebras (*Equus zebra*), plains zebras, which includes quaggas (*Equus quagga*), and Grévy's zebras (*Equus grévyi*).[16]

Zebras occur from Ethiopia to South Africa in a variety of habitats from treeless grassland to savanna to open woodlands. They show a slight sexual dimorphism in body size, with stallions usually being a little longer, taller, and heavier than mares. Stallions have thicker necks and possess canine teeth that are used to bite predators and other stallions when fighting; canine teeth are reduced or absent in mares.

Zebras have a black dorsal stripe running from the mane to the tail and a black ventral stripe. Grévy's zebras and mountain zebras have black and white stripes over most of their bodies, although they are broader and fewer in the latter (Figure 1.2). Plains zebras have the fewest stripes, and these are variable in distribution and appearance. Striping characteristics are prominent in the criteria used to identify zebra species, and each animal has distinctive striping that enables even humans to recognize individual zebras. Striped, upright manes and a long tail that terminates in a brush of long hairs are other characteristic features.[17]

Zebras prefer to eat grass but will also browse some herbaceous plants and even parts of woody plants and geophytes (rhizomes and corms) if food is scarce. Their incisors crop vegetation and their high crowned premolar and molar teeth have hard surfaces to chew tough plant

Figure 1.2. A Hartmann's mountain zebra and a plains zebra at left photographed in Etosha National Park, Namibia. Photograph by Yathin S Krishnappa. Creative Commons Attribution-Share Alike 4.0 International license. (A black and white version of this figure will appear in some formats. For the color version, please refer to the plate section.)

material.[18] They benefit from microorganisms living in their hindguts that ferment food and so release nutrients that otherwise would not be available. Zebras can live on vegetation with a low nutrient content, but they need to consume a lot of this food in order to survive and consequently they spend much of their time eating, depending on the quality of their forage.[19] One result is that often they need to graze even during the hottest parts of the day – unlike some grazing animals that can seek shade at this time.[20] Zebras usually drink water each day and will often walk long distances to reach it. When surface water is not available, they can dig holes to obtain it.

A zebra communicates by making sounds, by body posture, by facial expressions, and by movement of its head, mouth, ears, and tail. Animals often stand close together and groom and sniff each other. Taken together, these signals and interactions can communicate a whole range of messages: greeting, challenging, submission, affection, or readiness to mate. Communication and behavior are important in their social relationships and in surviving predators. Usually gestures of aggression are explicit enough that zebras can avoid fighting, but if not – as, for example, when a stallion challenges a dominant stallion for possession of his breeding group – animals may bite, rear up, and kick.

Stallions are attracted to mares during estrus and will attempt to mount them. If pregnancy results, a single foal is born after a gestation period of approximately twelve to thirteen months.[21] Foals have longer, softer, and browner hair than older animals. Befitting an animal born in open country with predators present, a newborn foal, whose legs are almost as long as those of adults, can stand and walk within an hour of birth and can run shortly thereafter. A mare licks and grooms her foal who becomes familiar with the mare's appearance, smell, and vocalizations. During this imprinting period, lasting for several days, a mare will not allow others to approach her foal. A mare will suckle her foal until weaning at between eight and thirteen months of age. Foals supplement milk with grazing when only a few weeks old. A mare may enter estrus within a month of giving birth and so foals may be born annually. Fillies and colts usually leave their mothers when they are one to two years old and become sexually mature by three to five years of age. Colts usually join stallion groups that provide some protection against predators and where the animals have play fights which prepare them for later combat with stallions.

Plains zebras and mountain zebras form breeding groups consisting of a stallion, one or more mares and their foals. Members of a breeding group care for each other, especially by protecting foals when predators are present, and remain together even if the stallion dies or is replaced. Within the breeding group, mutual grooming is an important activity that strengthens social bonds: it occurs between mares and foals, between stallions and mares, and between stallions, but very rarely between mares. After a foal has imprinted on its mare, other members of the breeding group groom the new arrival – establishing a social bond among all. Breeding groups stay together while grazing and walking to find water or vegetation. Often, several breeding groups merge to form a herd (now sometimes called a "zeal" or a "dazzle"). Grévy's zebras do not form breeding groups but have a fission–fusion social organization in which stallions defend their territories but may also at times share these with conspecifics, which may include stallions, mares, and foals; this association provides some protection against lions.[22]

Predators of zebras include spotted hyenas (*Crocuta crocuta*), lions (*Panthera leo*), leopards (*Panthera pardus*), cheetahs (*Acinonyx jubatus*), African wild dogs (*Lycaon pictus*), and crocodiles (*Crocodylus niloticus*). Wildebeests (gnus), antelopes, and ostriches that graze with zebras often provide warning that predators are approaching.[23] When attacked, zebras fight ferociously: the hard, sharp edges of their hooves can deliver powerful blows and their bites are damaging. Although zebras may sleep for up to seven hours a day, they often do so while standing to enable rapid flight and while some sleep one or more members of a herd will remain watchful.[24] The social structure of mountain zebras and plains zebras provides some protection against predation: an alarm cry from one will alert others. The animals will form an arc to face a predator that approaches too close, and a stallion will fight it. Members of the breeding group will also gather around a mare with a young foal to protect them.[25]

Besides predators, smaller organisms also take their toll: worms, flies, ticks, and lice all obtain nourishment from zebras. Particularly problematic are biting flies that annoy, draw blood, and may spread diseases; they are countered by stripes, skin odor, and tail swishing.[26] Microorganisms can be debilitating and deadly: bacteria, viruses, and protozoa cause anthrax, African horse sickness, and trypanosomiasis, respectively.[27] Yellow-billed oxpeckers (*Buphagus africanus*) and red-billed oxpeckers (*Buphagus erythrorhynchus*) that perch on zebras and eat their ectoparasites are a mixed benefit as they also take blood from sores on the animals.[28]

Mountain Zebras, *Equus zebra*[29]

As their vernacular name indicates, these animals live in mountainous areas and at one time must have formed a single population extending in a long swath in the west of southern Africa.[30] However, they now exist as two subspecies in geographically isolated populations: Cape mountain zebras (*Equus zebra zebra*) in western South Africa and Hartmann's mountain zebras (*Equus zebra hartmannae*) in Namibia.[31] Stallions of Hartmann's mountain zebras have an approximate shoulder height of 1,500 mm (59.1 inches) and an average weight of 298 kilograms (657.0 pounds); they are larger and heavier than Cape mountain zebras.[32]

Mountain zebras are the only striped equines with a dewlap, loose skin hanging from their throats; they also differ from other zebras in having a "grid-iron" or "fish bone" pattern of stripes on top of their rumps.[33] They are striped all over their bodies, but stripes do not extend to the ventral stripe and so their bellies are white, except for the ventral stripe. In comparison with plains zebras, they have more stripes, and these are narrower, except on the rumps (Figure 1.2). Their ears are larger and rounder than plains zebras, their hooves smaller, and only occasionally do they make a barking call similar to *kwa-haa*; this sound seems to serve as an alarm call. Their hard hooves are well suited to travel over the rocky ground of mountainous areas but during the rainy season they live on the plains and may even enter desert and coastal areas.[34]

Births occur throughout the year, with a peak in the late spring and early summer.[35] Young foals eat their mother's feces which provides a ready means of acquiring needed gut microorganisms.[36] Foals are weaned before they are one year old and leave the breeding group before they are two years old. Fillies may form nonbreeding groups, and colts do likewise, though their groups may be joined by stallions that have failed to secure mares, or that have been ousted from their breeding groups. At maturity, stallions may form a new breeding group with a filly from a nonbreeding group or challenge a stallion for his mares.

Cape mountain zebras are a conservation success story: their numbers had decreased to less than a hundred in the mid-twentieth century, but recent estimates show 1,714 mature individuals distributed on several reserves in South Africa, and the IUCN (The International Union for Conservation of Nature) lists them as being of "Least Concern."[37] Hartmann's mountain zebras, with a population estimated to be 33,265 mature individuals, are listed as "Vulnerable" by the IUCN; this designation probably reflects that they are still hunted and that large numbers could be lost if there were a calamitous drought.[38]

Plains Zebras, *Equus quagga*[39]

In 1825, the zoologist John Edward Gray (1800–75) named the plains zebra *Asinus burchellii*. He placed this zebra in a different genus, the *Asinus* of wild asses rather than the *Equus* of horses. The species name he took from the explorer and biologist, William John Burchell (1781–1863), who had supplied the British Museum with skins of the animal.[40] As the genus name *Equus* had precedence, the plains zebra was later given the binomial name *Equus burchellii*. DNA research, however, indicates that quaggas and plains zebras are the same species, with the species name *quagga* having precedence, and so the scientific name for all plains zebras is now *Equus quagga*.[41] The evidence for this conclusion and its implications for rebreeding are important and will be discussed in Chapter 7.

The evolutionary history of plains zebras has produced animals of differing appearance extending over a wide area. About 367,000 years ago, plains zebras possibly living in the Zambezi-Okavango wetland areas (present-day Botswana and Zambia) expanded north to the Equator and south to the tip of southern Africa.[42] Over this distance and time, morphological differences developed so that plains zebras vary in size and degree of striping, with both these characteristics forming a cline – a gradient over their range. Generally, the smallest most-striped animals occur in equatorial regions and the larger, least-striped animals are present at the southern end of their range. The smaller size is probably an adaptation to heat as the increased surface-area-to-volume ratio allows better cooling, and the more marked striping protects against biting flies.

Taxonomists had a hard time sorting out plains zebras and many taxa were described. Some plains zebras have distinct black and white stripes, and others have less distinct striping and brown "shadow stripes" within their white stripes. Some animals have stripes over their whole bodies, including their legs, and others have few or no stripes on their legs. Sometimes body stripes extend all the way to the ventral stripe and sometimes they do not and so the belly is white, apart from the ventral stripe (Figure 1.2). In 1936, Angel Cabrera, who had noted variation in coat coloration and striping within a single herd of plains zebras, realized that biologists had named approximately twenty different species and subspecies based on minor variations that occurred between individuals within the same population.[43] Reviewing the many descriptions of plains zebras, Cabrera reduced the number of subspecies to four; these did not include quaggas, which he viewed as a different species.[44]

More recently, six subspecies have been recognized: *Equus quagga boehmi* (Boehm's or Grant's zebra), *Equus quagga borensis* (Maneless zebra), *Equus quagga burchellii* (Burchell's subspecies), *Equus quagga chapmani* (Chapman's zebra), *Equus quagga crawshayi* (Crawshay's zebra), and *Equus quagga quagga* (quaggas).[45] Rau, however, questioned whether it is even appropriate to divide up plains zebras into subspecies at all.[46] This point of view is supported by studies of mitochondrial DNA.[47] A recent report argues against recognizing subspecies based on morphology and instead identifies ten populations based on their genomics and location.[48] Two of these populations in South Africa correspond to the subspecies accepted by most taxonomists: quaggas and *Equus quagga burchellii*; the latter is designated as Burchell's subspecies throughout this book. The name "Burchell's subspecies" should not be confused with the term "Burchell's zebra," which was used to denote all plains zebras when their binomial name was *Equus burchellii*; the name "Burchell's zebra" is outdated and so appears in this book only when it is part of a quotation.

Sizes of plains zebras are known from measurements taken in the Serengeti (northern Tanzania) and in Kruger National Park (South

Africa).[49] Adult stallions of *Equus quagga burchellii* in Kruger National Park had a mean body length of 2,800 mm (110.2 inches) measured from the first upper incisor to the end of the fleshy part of the tail, a mean shoulder height of 1,363 mm (53.7 inches) and a mean body mass of 318.5 kilograms (702.2 pounds); values for nonpregnant mares were essentially the same, at a mean length of 2,794 mm (110.0 inches), a mean shoulder height of 1,345 mm (53.0 inches), and a mean body mass of 321.6 kg (709 pounds). Plains zebras in the Serengeti (the subspecies *Equus quagga boehmi*) were considerably smaller: adult stallions had a mean body length of 2,521 mm (99.3 inches), a mean shoulder height of 1,277 mm (50.3 inches) and a mean body mass of 247.8 kilograms (546.3 pounds) and although their mares had a similar mean body length of 2,513 mm (98.9 inches), they had a lesser mean shoulder height of 1,227 mm (48.3 inches) and a lesser mean body mass of 219.1 kg (483.0 pounds).

Some stallions lead a solitary existence, others live together in stallion groups, and those animals that have prevailed over competitors live as part of a breeding group. Mares come into estrus each month and display their readiness to mate, which they may do several times a day.[50] For plains zebras in South Africa, both mating and foaling usually occur in the rainy season from October to March; pregnancy lasts about twelve months.[51] Mares may enter estrus shortly after foaling and so sometimes foals are born at approximately yearly intervals. A newborn foal weighs on average 33.7 kilograms (74.3 pounds) and has a mean shoulder height of 889 mm (35.0 inches).[52] A foal can stand and walk within an hour after birth and shortly thereafter it begins to drink its mother's milk.[53] At several weeks of age, foals begin grazing but continue to feed from the mother until they are about eleven months old.

Young stallions leave their breeding group between the ages of one and four years to join a stallion group. The members of a stallion group stay together – often within a much larger herd. They interact with each other until they are mature at about four to five years old. At this time, a stallion might successfully challenge an older stallion for his breeding group or abduct a filly. Fillies depart from their breeding group between the ages of 1 and 2.5 years when their estrus attracts suitor stallions; they first give birth at three to four years. If the filly joins a new breeding group, her stallion protects her until she is accepted by the established mares, who are initially hostile to her.

As with other zebras, a plains zebra can communicate by the overall posture of its body, by moving its mouth, ears, and tail, and by vocalization. Frequent sounds include warning snorts, a blowing sound to show satisfaction, and a kwa-haa barking sound that is often repeated; this cry differs slightly from animal to animal, which suggests that it may help animals recognize each other. Plains zebras also squeal and squeak – both sounds probably indicating that the animal is hurt or afraid.

Plains zebras inhabit a variety of habitats ranging from grassland with few or no trees to open woodlands; they are absent from dense woodlands and deserts. They occur at densities varying from less than one per square kilometer to just over twenty per square kilometer.[54] Their premolar and molar teeth, adapted to grazing abrasive silica-rich grasses, can chew tough plant material from other sources. In comparison with ruminants such as wildebeests that often graze alongside them, plains zebras are able to live on less nutritious vegetation. Typically, their food has a low nutrient content and so they have to spend much of their time eating.[55] One study showed that, between sunrise and sunset, lactating females grazed for 63.3 percent of the time, breeding group males for 63.5 percent of the time, and bachelor group males for 46.8 percent of the time.[56] Although they spend much of their time grazing, sleeping, resting, grooming each other and walking to water, plains zebras also roll in the dust, rub themselves against trees and rocks, and interact with their fellows in play.

Plains zebras frequently graze with ostriches (*Struthio camelus*), wildebeests (*Connochaetes gnou* and *C. taurinus*), and springboks (*Antidorcas marsupialis*). Plains zebras and wildebeests may even eat the same plants – with the wildebeests feeding on the leaves and leaving the more fibrous stems for the zebras.[57] Collectively, plains zebras, ostriches, springboks, and wildebeests warn each other of the approach of predators by their cries and their evasive movements.

When visiting a water source, plains zebras drink on average 4.7 liters (1.24 gallons).[58] Although it is often stated that they need to drink water each day, drinking every second – or even third – day has been reported.[59] Plains zebras usually stay within 12 kilometers (7.5 miles) of a water source and will dig holes to obtain it if supplies are short.[60] They often spend much time walking to water, with a lactating mare often leading her breeding group in this move.[61] Several breeding groups will travel together as a herd, and larger herds may number several thousand animals. Plains zebras may range over remarkably long distances with migrations of 500 kilometers (over 310 miles) having been reported and with some animals traveling more than 50 kilometers (31 miles) per day.[62]

Crocodiles kill plains zebras at waterholes, but lions inflict the heaviest toll. Other predators include spotted hyenas, leopards, cheetahs, and African wild dogs. Captain William Cornwallis Harris (1807–1848), a British military engineer who hunted in South Africa, noted the responses of plains zebras to danger:

> The senses of sight, hearing, and smell, are extremely delicate. The slightest noise or motion, no less than the appearance of any object that is unfamiliar, at once rivets their gaze, and causes them to stop and listen with the utmost attention – any taint in the air, equally attracting their olfactory organs. Instinct having taught these beautiful animals that in union consists their strength, they combine in a compact group when menaced by an attack either from man or beast; and if overtaken by the foe, they unite for mutual defence with their heads together in a close circular band, presenting their heels to the enemy, and dealing out kicks in equal force and abundance. Beset on all sides, or partially crippled, they rear on their hinder legs, fly at the adversary with jaws distended, and use both teeth and heels with the greatest freedom.[63]

Ectoparasites such as flies, ticks, and lice vex plains zebras, obtain nourishment from them, and may also cause diseases. Intestinal worms are a problem, as are diseases such as babesiosis, brucellosis, trypanosomiasis, anthrax, and African horse sickness. In spite of predation and diseases, plains zebras can live up to twenty years in the wild and have lived longer in captivity.

Of the three species, plains zebras are by far the most numerous, hence their alternative name "common zebra." Nonetheless, their numbers are decreasing: in 2002 there were about 663,212 animals in the wild, but a 2016 estimate put their numbers at 150,000 to 250,000 mature animals.[64] Causes of this decrease include hunting, habitat loss, and climate change. The IUCN lists them as "Near Threatened," and their fragmented populations make them vulnerable to loss of genetic diversity. Loss of subspecies in particular locations is problematic, for example, the population of Selous zebras in Mozambique declined from approximately 20,000 to less than one hundred in the late twentieth century.[65]

Grévy's Zebras, *Equus grévyi*[66]

The third species of *Hippotigris*, Grévy's zebras, are recognizable by their large, rounded ears and by having many thin black and white stripes that are everywhere present except on the bellies and the upper hind legs near the base of the tail (Figure 1.3). They are the largest species of zebras, with a shoulder height of up to 1,600 mm (63.0 inches) and a body

Figure 1.3. Grevy's Zebra. Photograph by T. Bobosh. Creative Commons Attribution-Share Alike 2.0 Generic license. (A black and white version of this figure will appear in some formats. For the color version, please refer to the plate section.)

length of 2,500 to 2,750 mm (98.4 to 108.3 inches); adult stallions range in weight from 352.9 to 430.9 kilograms (771.6 to 992 pounds).[67]

Grévy's zebras had a wide range in Somalia, Ethiopia, and Kenya, but now exist as discontinuous populations in southern Ethiopia and northern Kenya. They live in grassland or shrubland, with grasses as their preferred diet, although they are versatile and will eat other vegetation if grasses are in short supply. They spend much of their time feeding and will range over an extensive area in search of forage and water – which they will dig for in dry riverbeds. As an adaptation to aridity, they can live longer without water than other species of zebras: five days in the case of stallions, but just two days for lactating females. Animals will walk for 10 to 15 km (6 to 9 miles) in search of water and forage and may range over a considerable area.

The social organization of Grévy's zebras differs from the breeding groups of other zebras. Some males live in stallion herds of two to six animals and others are more solitary and set up their territory marked out with urine or dung. However, these animals allow mares and even submissive stallions to enter their territory – although the latter are kept out if an estrous mare is present. Conflicts between stallions over territory or an estrous mare often involve braying, dominance displays, submissive displays, kicking, and biting. Mares become mature at three to four years and stallions at about six years. As a mare traverses the territory of a stallion in search of water and forage, her urination may attract his attention. If she is in estrus, the stallion's dominance displays and braying may be followed by courtship and copulation, but the pair do not stay together. After a gestation period of approximately 13 months, a single foal is born, usually during the rainy season.

A mare lives with her foal or most recent two foals, and several mares with their foals may group together to form a small herd. The mother suckles, caresses, and protects the foal which she weans by about one year, but which may stay with her until about two years old. After colts leave their mothers, they live in stallion groups or lead a solitary existence. Fillies join other fillies and later will have foals. The fission–fusion social structure of Grévy's zebras differs from the breeding groups of mountain zebras and plains zebras.

The numbers of Grévy's zebras have declined sharply from approximately 15,600 forty years ago to about 2,700 in 2016, and the IUCN lists their status as "Endangered."[68] Threats to their survival include predators, diseases, subsistence hunting, and lack of water and grazing which are exacerbated by climate change and competition with the domestic animals of pastoralists. The challenges of conservation are compounded when Grévy's zebras are the preferential prey of lions, which must also be conserved.[69] An additional problem is that Grévy's zebras can hybridize with plains zebras whose range overlaps with theirs.[70] The production of fertile hybrid offspring threatens the integrity of their gene pool, which is also at risk from inbreeding because of their low numbers in fragmented populations. Once hunted, Grévy's zebras are now legally protected and, since 2007, the Grevy's Zebra Trust has worked hard and imaginatively for their conservation.[71]

Sorting Out Zebras in Southern African History

The diversity of South Africa is evident in its people, its landscapes, and its biota. This richness extended to zebras, whose likenesses were captured by artists, beginning with Bushmen and whose descriptions occur in historic documents. |Xam had different words for mountain zebras and quaggas but Burchell recorded that Khoekhoe used the same word for both species.[72] References to quaggas or kwaggas might indicate

Equus quagga quagga or one of the other subspecies of *Equus quagga*, or even *Equus zebra*.[73] Most people do not ponder the finer points of animal taxonomy: distinguishing between zebras based on their degree of striping or on the presence or absence of a dewlap could hardly have seemed important to those eighteenth- or nineteenth-century observers who lumped together all zebras as quaggas or kwaggas.[74] Fortunately, some historic accounts refer to mountain zebras as "bergkwagga" and use the name "bontkwagga" or "bonte quagga" (from "bonte" or "bonti," meaning pied or striped) to distinguish striped plains zebras from the less-striped quaggas.[75]

Traveling from Cape Town to the interior, nineteenth-century explorers, hunters, and naturalists saw mountain zebras, quaggas, and Burchell's subspecies.[76] Their all-too-brief accounts provide us with some details of these animals, but much is lacking. Some gaps have been filled by archeologists, whose analysis of bones excavated in caves have provided details of the past lives of zebras. Most remarkably, molecular biologists can unravel their history using the DNA from living cells or ensconced in the hides and bones of their dead. As a result, our understanding of these animals is more substantial than it was just a few years ago. Applied appropriately, this knowledge may aid in the conservation of those zebras that remain.

2 · Quaggas

Afar in the desert I love to ride,
With the silent Bush-boy alone by my side:
O'er the brown Karroo, where the bleating cry
Of the springbok's fawn sounds plaintively;
And the timorous quagga's shrill whistling neigh
Is heard by the fountain at twilight grey;
Where the zebra wantonly tosses his mane,
With wild hoof scouring the desolate plain . . .
Thomas Pringle[1]

Before the species went extinct, two individuals were photographed. These photographs are important in themselves and in order to assess the depiction of quaggas in paintings and texts. There are five photographs of a quagga mare living at the London Zoo from 1851 until her death in 1872 (Figure 5.1). Stripes confined to the mare's face, neck, mane, and forebody give way to dark coloration on her hindquarters and her right side where the white of her belly is scarcely visible; her legs are unstriped and there is a "corn" on the inside of her left front leg.[2]

It is often stated that this mare was the only quagga to have been photographed; however, research for this book suggests that Gustav Theodor Fritsch (1838–1927) photographed another one in 1864. Fritsch traveled by ox wagon through southern Africa in 1863–66 transporting a tripod camera, chemicals, and glass plates, which he developed in a tent.[3] He visited the farm of Andrew Bain at Quaggafontein (Quagga Fountain), south-west of Bloemfontein in the Orange Free State and took photographs of wild animals living there; he seems to have combined the images and then reproduced the composite as a woodcut (Figure 2.1). Fritsch used a technique termed "photoxylography" to engrave photographs on to wooden blocks that were then used to make woodcut prints. Although the animals were photographed, their surroundings are less detailed and appear to have been drawn.[4]

Fig. 25. Quagga, junges Wilde-Beest und Blesbok.

Figure 2.1. "Quagga, young wildebeest, and blesbok." Woodcut made by photoxylography of animals that Fritsch photographed in April 1864. Figure 25 in Fritsch, Drei Jahre in Süd-Afrika, 1868

The dorsal stripe, flanked on either side by lighter lines, is conspicuous on the quagga's rump and extends to the dock of the tail.

Beginning with these images focuses us on the animals people were looking at, interacting with, describing, and depicting. The historic representations of quaggas show the ways people thought about the animals as well as what they looked like.

Descriptions and Depictions

The many descriptions and depictions of quaggas made in Africa and Europe reflect both the appearance of individual animals and the different ways that observers thought about these animals. For the indigenous people who hunted them, they were part of their livelihood and cultural life, whereas in Europe they were viewed as exotic animals. Some people perceived them as horses, others as donkeys, and these distinctions affected how artists portrayed them. Appendix 1 reviews illustrations of quaggas made prior to their extinction. Presumably, most of these images were made from living animals, or from animals that had been

Figure 2.2. Petroglyph of a quagga incised on andesite rock, 1000–2000 BP.
Image id: 01612928374. Courtesy of the British Museum. (A black and white
version of this figure will appear in some formats. For the color version, please refer
to the plate section.)

killed in order to illustrate them, although possibly some were copies of
earlier representations.

The indigenous people of southern Africa recorded quaggas' appearance in their rock paintings and petroglyphs. The explorer John Barrow
(1764–1848), who saw these depictions of zebras, quaggas, baboons,
ostriches, and several species of antelopes drawn with mineral ochers
and charcoal on the smooth walls of a cavern, admiringly described,
"The force and spirit of drawings, given to them by bold touches
judiciously applied, and by the effect of light and shadow."[5] These
images of quaggas (Figure 2.2), some of which are thousands of years
old, are readily distinguishable from fully striped zebras. What we know
of southern African languages indicates that some distinguished between
quaggas and other zebras. As noted in Chapter 1, the |Xam, the first

known inhabitants of the Karoo, called quaggas "‖Ƨkhwĩ", whereas they used "‖kabba" to denote mountain zebras.[6] Two other groups of indigenous people also knew quaggas: the Khoekhoe, who named them "quacha" and the Xhosa, who called them "iqwarha."[7]

Venturing inland from Cape Town in 1660, the colonists encountered zebras which they viewed through the lens of their own experiences in Jan van Riebeeck's words, "like the best horse one could wish for."[8] However, they failed to classify them into consistent categories and so the terms "quagga" or "kwagga" were applied by some people to both the well-striped mountain zebras and quaggas with their chestnut color and smudged out posterior stripes. Anders Sparrman (1748–1820), a Swedish biologist who lived in South Africa during 1772, 1775, and 1776, clearly appreciated the distinction between quaggas and mountain zebras. Befitting a student of Linnaeus, his brief 1786 account includes a name, vernacular knowledge, a description, and a comparison with a closely related species:

It was here that I saw for the first time in my life, one of the animals called *quaggas* by the [Khoekhoe] and colonists. It is a species of wild horse, very like the *zebra*; the difference consisting in this, that the quagga has shorter ears, and that it has no stripes on its fore legs, loins, or any of its hind parts.[9]

However, most Europeans only slowly worked through the puzzle of how quaggas fitted in with other zebras, even other equines: were they plains zebras or a separate species, and were they more horse-like or donkey-like? These were the questions that newcomers brought to their encounters with quaggas.

Robert Jacob Gordon (1743–95), who explored the interior of the Cape Colony in the late 1770s, was the first European to paint a quagga in Africa. On horseback, he pursued a foal and separated her from the rest of the herd, which numbered about seventy:

Seeing a herd of quaggas we gave chase to them, but they ran too fast. But a young one breaking away from the herd, we chased up close to it and the poor animal, which was about a month old, ran with the horses (without our haltering it) to the outspan, calling for its mother with a yelp that sounded like a jackal's.[10]

Gordon depicted the quagga foal aged about one month on November 23, 1777 (Figure 2.3). Young plains zebra foals have soft hair which is browner than in older animals and this is the case for the quagga

Figure 2.3. Een jonge kwagga. Female quagga foal depicted by Robert Gordon in 1777. Gordon, The Gordon African Collection, (RP-T-1914-17-190). Courtesy of the Rijksmuseum. (A black and white version of this figure will appear in some formats. For the color version, please refer to the plate section.)

foal in the Iziko South African Museum in Cape Town which has "long, woolly fur (25–30 mm long)."[11] This fluffy hair, which will be replaced later, is evident in Figure 2.3, including at the base of the tail where subsequently the dock will be present. Gordon's painting became influential in European imagery of quaggas: Allamand, Buffon, and Schreber copied it.[12]

Some quaggas were shipped to Europe, and they, too, were illustrated. These images document quaggas but also incorporate what some Europeans thought they were supposed to look like. The first notice to the scientific community about this new type of zebra appeared in *Gleanings of Natural History* by George Edwards (1694–1773), published in 1758. Edwards featured paintings of two zebras, a male mountain zebra and a female quagga (Figure 2.4) kept at the menagerie of the Prince of Wales at Kew Palace, London (now the Royal Botanic Gardens, Kew).

Figure 2.4. Zebra femina. George Edwards painted this quagga mare in 1751. Plate 223 in Edwards, Gleanings of Natural History, 1758. (A black and white version of this figure will appear in some formats. For the color version, please refer to the plate section.)

The two zebras had different striping, which Edwards interpreted as a case of sexual dimorphism rather than individuals of a mismatched set. He illustrated the male, a mountain zebra, with many black and white stripes over his legs and most of his body and described him as being "about the bigness of a mule, or a middle-sized saddle-horse: its general shape is like that of a well-made horse."[13] He described in detail the female, a quagga:

To speak of its general colour, (exclusive of its stripes, which are all black) the head, neck, upper part of the body and thighs, are of a bright-bay colour: its belly, legs, and the end of the tail, are white. On the joints of the legs it had such corns as we see in horses: the hoofs are blackish: the head is striped a little different from the last described [mountain zebra]: the mane is black and white: the ears are of a bay colour: it is a little white in the forehead: it hath several broad stripes round the neck, which become narrow on its under side: it hath a black list [stripe] along the ridge of the back, and part of the tail, and another along the middle of the belly: the stripes on the body proceed from the list [stripe] on the back, and some of them end in forks on the sides of the belly, others in single points, and these have some longish spots between them. The hinder part of the body is spotted in a more confused irregular manner. The two sides of this, as well as the last described [mountain zebra], were marked very uniformly.[14]

In his painting, the quagga had chestnut stripes in place of the white stripes of other zebras. At her rump the black stripes broke into black blotches which then disappeared altogether at the hindquarters to give a chestnut color similar to the color of the lighter stripes. Her mane showed the same banding as her neck – the white hairs aligned with the chestnut-colored hairs on her neck. Edwards did not describe the tail but painted it with specific detail (Figure 2.4). It comprises a "dock" – where the skin is covered only by short hairs – and a "skirt" or "brush" at the end, comprising a tuft of long hairs. This arrangement is like that in asses (donkeys) and is often termed an "asinine" tail. In horses, long hairs are present over the whole tail so that the dock is not visible.

When Edwards's mistaken classification was realized, the male and female animals were divorced taxonomically and given separate binomial names. The painting of the male became the type specimen for mountain zebras, and the painting of the female became the type specimen for quaggas. Other taxonomic questions had to do with less-closely related equines. Did quaggas have more in common with horses or donkeys?

Samuel Daniell (1775–1811), who was both secretary and artist for the 1801–02 Truter-Somerville expedition from Cape Town to north of the Orange River and who left a rich pictorial record of the journey, endorsed the connection with horses. His aquatint of an African scene with quaggas, "The Quahkah" (Figure 2.5) was the second image of the animal in its natural habitat to be made by a European, after Gordon's illustration a few decades earlier. It is overly horsey in several respects. The animal that Daniell recorded in the foreground was as muscular as the draft-horses that European farmers had selected through centuries of breeding.[15] The erect striped mane and brown upper body are like those in other quaggas, but other aspects differ. The dark stripes originating at the dorsal line extend down the lateral surface for only a short distance, unlike the stripes photographed in Figure 2.1, which extend over the entire lateral surface. The white of the belly and legs extends upwards so that more than half the lateral surface is white. This representation differs from other illustrations and taxidermy specimens where the dark stripes

Figure 2.5. The Quahkah. Plate 15 in Daniell, African Scenery and Animals, 1804–1805. Courtesy of the British Museum. (A black and white version of this figure will appear in some formats. For the color version, please refer to the plate section.)

and the surrounding brown of the back extend over most of the lateral surface so that the white of the belly is scarcely visible, as in Figure 5.1.[16] As for the tail, Daniell depicted it as horse-like, with a full skirt that extends almost to the body, making it difficult to discern whether a dock is present. This animal so dominates the scene that it is easy to overlook the two other quaggas in the middle ground near to the end of its tail. The closest of these is less stocky and is also horse-like in a different way – having longer legs in proportion to body length than quaggas. Daniell's fellow explorer, Dr. William Somerville (1771–1860), a Scottish physician, recorded a contrary description: "the tail only bushy at the end like an asses [sic]."[17] These two people observed the same quaggas but perceived the tails differently. In Daniell's defense, his original depiction may have been more accurate but might have been changed when his field drawings were produced as an aquatint in England.[18] In any event, *The Quahkah*, although frequently reproduced, is inaccurate in several respects.

Figure 2.6. Een Quagga. Aert Schouman painted this watercolor of a quagga stallion in 1780. Courtesy of the Teylers Museum. (A black and white version of this figure will appear in some formats. For the color version, please refer to the plate section.)

Daniell's *The Quahkah* stands in sharp contrast to the animal painted by Aert Schouman (1710–92) whose ears are long and whose mane is short and scraggy (Figure 2.6). Uncertainty clouds the expression on the bowed face of this Eeyore of quaggas.[19] The dissimilarity between this animal and the one that was painted next could hardly be more pronounced.

In 1793, Nicolas Maréchal (1753–1803) painted a stallion (Figure 2.7) that lived at the menagerie of King Louis XVI at Versailles, France.[20] The handsome animal in this painting has a full tail with a marked twist; his face, neck, mane, and forebody have dark reddish-brown stripes alternating with lighter reddish-brown stripes. At the mid-body, the contrast between the stripes lessens and they disappear completely at the hind body where the upper part of the body has a light reddish-brown coloration which gives way to the white of the unstriped legs and belly. It is a magnificent animal, but the painting is an eighteenth-century equine version of a flattering photoshopped magazine image: in eighteenth-century Europe, the equine worth celebrating looked a lot

Figure 2.7. Le Couagga. Nicolas Maréchal painted the quagga stallion of King Louis XVI in 1793. https://en.wikipedia.org/wiki/Quagga#/media/File:Quagga.jpg. (A black and white version of this figure will appear in some formats. For the color version, please refer to the plate section.)

like a horse. Of all the attempts to make the quagga horse-like, this was extreme, and Maréchal's liberties in his portrayal become clear from examination of the animal itself preserved as a taxidermy specimen in Paris. I have scaled a photograph of this specimen with a reproduction of the painting. This comparison shows that the artist exaggerated the length of the legs to make the animal seem more horse-like. This impression was further heightened by the dock-less tail being longer and fuller than in photographs (Figure 2.1), most paintings of quaggas, and certainly than in taxidermy specimens.[21] Maréchal's painting reflects the power of established ideas in obscuring the form of the animal. Artistic license, to depict the animal according to conventional equine beauty, was in play, but there was also a political imperative: this animal lived in King Louis's menagerie at Versailles and the royal quagga had to be a fine horse, not a common donkey.

Other artists were more ambivalent about whether quaggas more resembled horses or donkeys. Benjamin Waterhouse Hawkins (1807–94) made a lithograph of two quaggas on Lord Derby's Knowsley estate near Liverpool, England that showed one with a tail

Figure 2.8. *Quagga, Asinus quagga.* Waterhouse Hawkins painted these quaggas at the Knowsley Menagerie of Lord Derby. Plate 54 in Gray et al., Gleanings from Knowsley Hall, 1850. (A black and white version of this figure will appear in some formats. For the color version, please refer to the plate section.)

having a long dock and a short brush and the other having a fuller tail (Figure 2.8).[22]

These equine-like paintings of Daniell and Maréchal can be contrasted with those of the noted artist of animal subjects, Jacques-Laurent Agasse (1767–1849) that made a case for beauty that was less horse-like. Agasse was a renowned painter of horses and had a professional interest in showing the particular characteristics of this new form of equine. In 1817, he painted one of the quagga stallions that served the very horse-like function of pulling the carriage of Sheriff Joseph Parkins through Hyde Park. He followed this work in 1821 with a painting of another quagga stallion (Figure 2.9). Both paintings show animals with erect manes of black and white hairs, dark hooves, and striping that ends before their rumps.[23] Their tails have docks followed by skirts with hair of admirable, but not flowing, length. Standing alertly against a British landscape, the animal Agasse gives us is beautiful, but is not a horse.

The hunter and author William Cornwallis Harris (1807–48) provided a detailed description of quaggas, based on those he saw and shot on

Figure 2.9. A Male Quagga from Africa, the First Sire. Jacques-Laurent Agasse painted Lord Morton's stallion in 1821. Object number RCSSC/P 278. Courtesy of the Royal College of Surgeons. (A black and white version of this figure will appear in some formats. For the color version, please refer to the plate section.)

hunting expeditions in 1836 and 1837. Without the technology of photography, he provided a snapshot in words:

Adult male stands four feet six inches high at the wither [shoulder], and measures eight feet six inches in extreme length. Form compact. Barrel round. Limbs robust, clean and sinewy. Head light and bony; of a bay [reddish brown] colour, covered on the forehead and temples with longitudinal, and on the cheeks, with narrow transversal stripes, forming linear triangular figures between the eyes and mouth. Muzzle black. Ears and tail strictly equine; the latter white, and flowing below the hocks [ankles]. Crest very high, arched, and surmounted by a full standing mane, which appears as though it had been hogged [clipped], and is banded alternately brown and white. Colour of the neck and upper parts of the body, dark rufous [reddish] brown, becoming gradually more fulvous [orange-brown], and fading off to white behind and beneath; the upper portions banded and brindled [streaked] with dark brown stripes, stronger, broader, and more

Figure 2.10. Equus quagga. The Quagga. Detail of Plate 2 in Harris, Portraits of the Game, 1840. (A black and white version of this figure will appear in some formats. For the color version, please refer to the plate section.)

regular, on the neck; but gradually waxing fainter, until lost behind the shoulder in spots and blotches. Dorsal line dark and broad, widening over the crupper [rump]. Legs white, with bare spots inside above the knees.[24]

Harris concluded this description: "Female precisely similar. Has an udder with four mammae."[25] The last sentence suggests that Harris was not always a reliable observer: taxidermy specimens show that quaggas had two nipples – as do other plains zebras – yet he claimed that quaggas and other plains zebras had four nipples. Harris recorded shooting "many" quaggas, so how could he fail to count an obvious feature correctly?[26]

Harris painted quaggas in their Karoo habitat (Figure 2.10). Despite the naturalism, it raises additional questions about the quality of his observations. It shows in the foreground an animal that is strikingly similar to the one painted by Daniell (Figure 2.5) in overall appearance, stance, and full horse-like tail –the latter feature shared with the quaggas depicted in the background. The unnaturally short dark stripes on the neck and forebody resemble those of Daniell's aquatint, as does the white of the belly extending approximately halfway up the body and rump – features not present in taxidermy specimens. Undoubtedly Harris copied his quaggas from Daniell's painting, but it is unclear why he would have done so when he himself had seen so many of the animals.[27] Possibly, he delegated production of this figure to someone who was influenced by Daniell's aquatint. Harris portrayed the tail longer than Daniell had, to accord with his own description, "flowing below the hocks [ankles]." The tail was beautiful, but most taxidermy specimens and contemporary illustrations do not show such length.[28]

There were other problems with the painting. The brown color of the quaggas painted by Harris was lighter than both the "dark rufous brown" of his own description and the color of Daniell's quagga; the wildlife writer Henry Anderson Bryden (1854–1937) also viewed Harris's coloration as "a trifle too light."[29] Other observers have noticed that Harris's color plate was "impossibly idealized," featuring "incorrect drawing and colouring," and "somewhat too equine."[30] The latter critique goes to the heart of Harris's vision of quaggas, which he described as, "Compact, strong, and muscular, with clean bony limbs, and a foot which might serve as a model to the veterinary student, this *petit cheval* cannot fail forcibly to remind us in all its form and proportions, of the horse in miniature."[31]

The question whether quaggas more closely resembled horses or donkeys became a debate, with Harris taking the horse side, in opposition to the famous comparative anatomist Baron Georges Cuvier (1769–1832) who, in 1827, had described quaggas in similar terms as Harris, except for their tails: "This species reminds us of the forms and proportions of the Horse, by the lightness of its figure, and the smallness of its head and ears, but it has the tail of the Ass."[32] Cuvier's assessment would have created a quandary for Harris: Cuvier's description of quaggas pulling a carriage like "well-trained horses" (see next quote) aligned with Harris's view of quaggas, but the asinine tail contradicted it – Harris believing that quaggas had horse-like tails.[33] Curiously, in refuting this esteemed scientist in 1840, Harris did not draw on his knowledge of living quaggas in the wild, but instead cited European, urban evidence, with a *non sequitur*:

Even the Baron Cuvier has fallen into the error of describing the Quagga to be the proprietor of an asinine tail – a mistake which is the more surprising, since it is stated by the same author in his "*Règne Animal*," that, "among the equipages occasionally exhibited in the gay season in Hyde Park, and other fashionable places of resort, may be seen a curricle drawn by two Couaggas, which seem as subservient to the curb and whip as any well-trained horses."[34]

Ipso facto quaggas had horse-like tails because they pulled the carriages of stylish people.

Others who followed Harris expressed similar opinions, although with less fervor: quaggas were said to be "more a horse than any of the living zebras," "more equine than asinine in character," "more horse-like even than Burchell's zebra," and, in almost identical words, "more like a true horse" than a plains zebra.[35] Clearly, the portrayals by Daniell and Harris influenced these opinions, but so did the European fashion of hitching quaggas to carriages.[36]

In contrast to their portrayals by Daniell, Harris, and Maréchal, most illustrations of quaggas show tails with distinct donkey-like docks. There was variation in tail morphology in quaggas, with some having longer and fuller skirts than others, but none appear to have the horse-like tails depicted by Daniell and Harris, and even if such tails had existed they would have been exceptional and not the norm as presented in Harris's account.[37] What we believe influences what we see, and this tendency certainly seems to be the case when Daniell, Harris, and Maréchal gave horse-like attributes to quaggas.[38] Commenting on a similar instance where pre-conceived ideas had influenced interpretation, the biologist Stephen J. Gould noted "the subtle and inevitable hold that theory exerts upon data and observation."[39]

Taxidermy Specimens

Besides illustrations and photographs, taxidermy specimens also tell us what quaggas looked like. These specimens and their museum locations are described in more detail in Chapter 7, but their dimensions are included here as they provide evidence for sexual dimorphism in body size, with the females being, on average, larger than the males.[40] The head–body length of the males ranges from 1,930 to 2,300 mm and averages 2,070 mm (81.50 inches), whereas the head–body length of the females ranges from 2,100 to 2,490 mm and averages 2,294 mm (90.32 inches). The specimens show the same sexual dimorphism in shoulder height: this measurement in males ranges from 1,090 to 1,200 mm and averages 1,140 mm (44.88 inches) while the shoulder height of the females ranges from 1,100 to 1,285 mm and averages 1,189 mm

(46.81 inches).[41] Statistical analysis shows that female specimens were significantly longer (by 224 mm or 8.82 inches) and marginally taller (by 49 mm or 1.93 inches) than the males.[42]

These measurements differ from Harris's account, which made no mention of sexual dimorphism in size and cited the shoulder heights of quaggas as 48–52 inches (1,219–1,321 mm).[43] The small number of specimens – particularly males – raises the problem of sampling error. Additionally, hides could have shrunk before mounting, especially those that were shipped from Africa to European museums, and this change might be responsible for most measurements being less than those recorded by Harris for recently shot animals; or perhaps he overestimated. Once at their destination, hides could have been distorted during mounting. However, it is difficult to see why artefacts created by shrinking and mounting would have given rise to the size differential between male and female specimens. Although the applicability of data from eighteen taxidermy specimens to historical quaggas is debatable, it appears that males were smaller than females: particularly striking is the 224 mm (8.82 inches) difference in average head–body length and that the three specimens with the shortest head–body lengths are all males. Craniometric studies provide confirmatory evidence – showing that females were larger in both quaggas and the closely related Burchell's subspecies.[44]

Most accounts of plains zebras state that males are larger. However, the differences were minor among plains zebras in Kruger National Park: stallions averaged 1,363 mm (53.66 inches) in shoulder height and mares averaged 1,345 mm (52.95 inches); stallions averaged 2,800 mm (110.23 inches) in total length and mares averaged 2,794 mm (110.00 inches). Mares were on average 3.1 kg (6.83 pounds) heavier.[45] Compared with these other southern African plains zebras, the sexual dimorphism in body size of quaggas seems less remarkable.[46] Many quaggas lived in arid environments, and the larger size of the females could have evolved to provide adequate food reserves to allow lactation and pregnancy during droughts when the vegetation was not providing enough nourishment. It is also possible that the existence of quaggas as an isolated population led to random genetic drift, resulting in sexual dimorphism. Scientists have evoked both random genetic drift and selection as likely reasons for the coat coloration of quaggas, and it is possible that one of these factors or a combination of both was responsible for females being larger than males. Whatever the explanation, the data dispute Harris's claim of "Female precisely similar" and the assertion by the explorer Thomas Baines

(1820–1875) that mares were "more delicately formed and lighter-coloured."[47]

Observations by the hunter Frederick Courteney Selous (1851–1917) of plains zebras from the interior of South Africa are relevant to this account of sexual dimorphism. He described these animals as being of shoulder height 52 inches (1,321 mm) or more and did not indicate height or length differences between mares and stallions. However, he noted that, "The mares become much heavier and more bulky than the stallions" because of "fat an inch thick ... between the skin and the flesh."[48] This fat was probably an adaptation to food scarcity in a closely related subspecies living under environmental conditions similar to those of quaggas, and likewise led to sexual dimorphism.

The preserved hides of taxidermy specimens show variations in color; for example, the Amsterdam, Berlin, and Paris quaggas are dark brown, whereas the Munich quagga is much lighter, probably because of fading.[49] The colors of other quagga hides vary in their shades of brown, but none have the "bright-bay colour" described by Edwards.[50] Rau suggested that coat coloration could have varied seasonally, or between quaggas, or because of fading in museum specimens.[51] Although these possibilities are reasonable, ultraviolet light causes fading in the coats of taxidermy specimens, and so this process seems to be the predominant reason that no quagga hides show the reddish color depicted in some paintings.[52]

Quaggas, like other plains zebras, had a dark dorsal stripe that extended from the mane along the length of the back and ended in the dock region of the tail; this stripe, which can be seen in Figure 2.1, is present on the hindquarters even when other stripes are absent. Quaggas also had a ventral stripe running along the middle of the abdomen, which was otherwise unstriped.

Quagga museum specimens and illustrations of living animals show alternate darker and lighter stripes on their faces, necks, and forequarters, but differ in the degree to which this striping is continued over the rest of the animal. In the Bamberg, Berlin, and Munich quaggas, distinct stripes end at the shoulders, whereas in the Mainz quagga mare they continue to the beginning of the rump. As noted by the Quagga Project, "It is evident from the 23 preserved skins of the extinct Quagga, that this former population displayed great individual variation."[53]

The quaggas with most stripes resemble the least-striped specimens of the other southern subspecies of plains zebras, Burchell's subspecies, which have more body stripes and most have some stripes on their legs.

The gradation in striping often makes it difficult to assign an animal to one subspecies or another.[54] The two subspecies intermingled near the eastern edge of the quagga's range in the Orange Free State.[55] These closely related animals would have almost certainly interbred and so eastern quaggas could have had more stripes than those that lived in the west where they would have been isolated from other plains zebras. However, Groves and Bell tentatively concluded that well-striped and less-striped quaggas occurred throughout their range, and even within the same herd.[56] The absence, in many cases, of records showing the origins of taxidermy specimens or of animals that were illustrated makes this issue impossible to resolve.

Classification

Based on George Edwards's painting and description, the Dutch naturalist Pieter Boddaert (1730–1795) gave quaggas their binomial name: *Equus quagga* and described the color of the upper body as brown ("fusco") and the lower body as white ("albo"). As the first person to give the animal a diagnosis (species description) and a binomial designation (Figure 2.11), Boddaert is the authority.[57] Johann Friedrich Gmelin (1748–1804) also described quaggas in 1788 and gave them the same name.[58] Scientists have cited either Boddaert or Gmelin as the authority

*Zebra. 5. E. corpore albido rufefcente fafciis transverfis. fufcis. LINN. 12. p. 102. 3. *Zebre.* BUFF. XII. tab. 1. 2. *Zebra.* PENN. *quadr.* p. 13. 4. EDW. *glean.* tab. 222. KNORR *delic. Select.* II. *tab.* K. 8.
Habitat ad *Cap. bon. Sp.*

†Quagga. 6. E. capite caudaque fafciis fufcis corpore fufco, fubtus albo. *Kwagga.* BUFF. *f uppl.* XI. p. 150. tab. 7. *The Zebra female.* EDW. *glean.* 223. *Quacha.* PENN. *quadr.* p. 14. 5.
Habitat in *Caffrorum* regione.

Figure 2.11. Boddaert's diagnoses of mountain zebras and quaggas. Boddaert, Sistens Quadrapedia, 1785.

for the name, *Equus quagga*, although properly this honor belongs to Boddaert as he published first.

The person who writes the diagnosis for a species usually puts a corresponding type specimen in a collection, for example, in a museum. This was not done in this instance and so the type specimen is Edwards's painting rather than a preserved animal.[59] Normally, the location cited in the diagnosis serves as the type locality for the species, but Boddaert's "Habitat in *Caffrorum* regione," is imprecise, and so the type locality for *Equus quagga* is the Seekoei River (west of Colesberg, Northern Cape Province) where Gordon described the female quagga foal illustrated in Figure 2.3.[60] The binomial name, authority, diagnosis, type specimen, and type locality are important information for biologists who want to identify a species precisely.

Biologists occasionally revise classifications, and people sometimes gave quaggas different names. Some changed the spelling of the species name to *qouagga*, *quaccha*, and *quacha*.[61] Even the genus was in question: Boddaert and Gmelin both put quaggas and zebras within the genus *Equus*, but John Gray placed quaggas with other plains zebras and mountain zebras in the same genus as asses, and named them *Asinus quagga*.[62]

Variations in coat coloration of quaggas led to the naming of several subspecies: *E. quagga danielli*, *E. quagga greyi*, *E. quagga lorenzi*, *E. quagga quagga*, and *E. quagga trouessarti*.[63] These trinomial names, which were proposed when quaggas were classified as a different species from plains zebras, are no longer valid and were probably conferred because taxonomists working in European museums knew animals only from isolated hides and so did not appreciate the variations in coloration that occurred among quaggas within the same population.

Charles Hamilton Smith (1776–1859) named a new genus for zebras, *Hippotigris* – using the word (tiger horse) that Romans had given to Grévy's zebras – and named two species, *H. isabellinus* and *H. quacha*. Judging from the description and painting, the latter was a quagga.[64] However, the name *Hippotigris quacha* does not have priority over *Equus quagga* and is not valid. The word *Hippotigris* has not been lost and denotes the subgenus of zebras within the genus *Equus*.[65]

Quagga Habitats

Quaggas ranged from just east of Cape Town to the grasslands of the Orange Free State and eastern parts of the Cape Colony, an area that included several biomes: the Nama Karoo, Succulent Karoo, Grassland, Savanna, Albany Thicket, and Fynbos biomes.[66] It is possible to speculate about the plants quaggas ate, based on what is known about other plains zebras which have grasses as a major component of their diets, but which will eat plants other than grasses and will even browse on roots and woody parts of plants if choicer fare is not available. It is reasonable to think quaggas did likewise, provided the vegetation did not contain toxic or distasteful substances. The premolar and molar teeth of quaggas in the skulls of museum specimens resemble those of other plains zebras, and the ability of these teeth to chew silica-rich grasses would have enabled quaggas to eat tough plant material.[67] Hindgut fermentation in quaggas would have extracted nutrients from this vegetation, just as it does in other plains zebras.

The principal habitat of quaggas was the Karoo, a region that occupies more than a third of the area of South Africa (Figure 2.12).[68] The

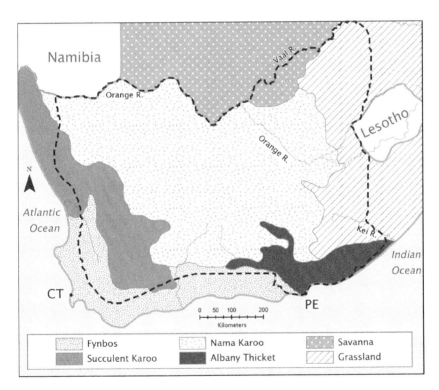

Figure 2.12. Biomes occurring in areas where quaggas lived (within the dashed line). This figure by the author was prepared by Bruce Boucek and Camille Tulloss; it is modified from a map published in Heywood 2015 and is reprinted courtesy of Kronos

northern part of the Karoo, the Nama Karoo, is a plateau at approximately 1,000 to 1,400 meters (3,300 to 4,600 feet) above sea level with a surface broken by ravines (kloofs) containing riverbeds dry for much of the year. Erosion of the surrounding rocks left behind hills (koppies) that rise from the plateau. The soil is usually shallow, and the ground is often rocky, with limestone being common. Henry Bryden described the habitat and its fauna enthusiastically, but by the time he wrote he was compelled to use the past tense.

This huge tableland, standing 4,000 feet above sea-level, was ... crowded with the most magnificent of fauna. The eland, the gemsbok, the hartebeest, the bontebok, wildebeest (gnu), quagga, blesbok, ostrich, and many of the small antelopes, all abounded in extravagant profusion ... The lion everywhere stalked in lordly pride; the ponderous rhinoceros roamed far and wide; and the elephant wandered freely across from one grazing-ground to another.[69]

Over geological time, global fluctuations in climate accompanying ice ages and interglacial periods would have affected populations of quaggas and other organisms living in the Karoo. Current temperatures are usually between 0°C to 40°C (32°F to 104°F), with frosts occurring during the winter, especially at higher altitudes.[70] Rainfall is variable across this region: it is below 100 mm (approximately 4 inches) per year at some places in the west, where most rain falls during the winter months of May to September, and up to 500 mm (approximately 19.7 inches) in the east, where most rain falls from October to April.[71] At the western part of the quagga's range, there was also a south–north gradient in rainfall, with lower rainfall in inland areas.[72] Bryden recorded how seasonal rains change the face of the Karoo: "Its rich red soil, though baked and sun-dried by the burning sun, is capable of supporting a luxuriant vegetation, and when the rains fall the withered shrubs and plants instantly are transformed into a blazing carpet of the most brilliant flowers."[73]

Rain would have brought not only flowers but drinking water and new vegetation, and so animals that had survived the dry period would build up food reserves against future scarcity.

Aridity was, however, a perennial problem in the Nama Karoo, and it was intensified by winds that increased evaporation and transpiration. Sometimes failure of even the sparse rain led to devastating droughts, as in this description from 1822:

At the time of our journey no rain had fallen on the Karroo for upwards of twelve months, so that I saw it under its most desolate aspect. Not a vestige of green pasturage was to be descried over the surface of the immense monotonous landscape; and the low heath-like shrubbery, apparently as sapless as a worn-out broom, was the only thing our cattle had to browse on. No wild game was to be seen: all had fled apparently to some more hospitable region. Not even a wandering ostrich or bird of prey appeared to break the death-like stillness of the waste.[74]

Such droughts occurring in the past would have affected both the plants and the populations of animals that depended on them.[75] Organisms living in the Karoo might have had adaptations to its cold and dry conditions, and some could have migrated to more favorable habitats when these circumstances prevailed.

Kloofs (ravines) were important for quaggas as protection from the wind, cold, and sun of the open Karoo, and this environment also provided sheltered habitats for plants such as the mimosa tree, *Vachellia karroo*. When the rare rains came, they were often heavy and created torrents in the kloofs. After the rains had ceased, pools left on the riverbeds provided water for quaggas during drier periods. Martin Hinrich Carl Lichtenstein (1780–1857), a physician who explored southern Africa from 1803 to 1806, provided this description of the rivers:

In those places only, where the rivers begin to flow into the plains, are they shaded with mimosas and other little trees. In the higher parts, the beds of the rivers are often broad, and composed of an argillaceous schist; the stones in many places form a sort of steps, down which, during the rainy season, cascades rush from step to step in a very picturesque manner. In the summer time when the streams cease to flow, the greater part of the springs are seen rising from the interstices of these steps.[76]

Upon leaving the plateau, kloofs broadened into valleys which would have provided pasture, water, and shelter for quaggas.

The predominant vegetation of the Nama Karoo comprises perennial grasses, dwarf shrubs, small trees, and annual plants;[77] all these plants could have sustained quaggas, provided that they were palatable.[78] However, the availability of water and vegetation would have affected the size and distribution of quagga populations.[79] During dry periods, the Nama Karoo could not have sustained significant populations of large herbivores such as elephants, rhinoceroses, antelopes, mountain zebras, and quaggas, and so these animals would have suffered from thirst and hunger, or would have migrated in search of food and water.[80]

Some of these migrations were spectacular: in 1849, quaggas were part of a migration of springboks, wildebeests, blesboks, and elands that thronged the gardens and streets of Beaufort West for three days and left the veld stripped of vegetation.[81] The next year, "wild animals of every description, forced from their more remote haunts by the continued drought but ready to return the moment their instinct should discover that rain had fallen here" crowded into an area near Bloemfontein "so that people could not see the ground."[82] Remarkable as these sights must have been, the toll of large-scale hunting prevented their repetition.

Quaggas occurred in grassland of the Orange Free State and the eastern parts of the Cape Colony. Parts of this region receive 500–600 mm (approximately 20–24 inches) of rain per year, and the soil is productive, so its extensive grasslands could have sustained large populations of herbivores, including quaggas, as reported by Baines in 1850.[83] The northeastern part of this area, the Highveld of the Orange Free State, a plateau at an altitude of approximately 1300 meters (approximately 4,300 feet) above sea level, provided a refuge for quaggas into the 1870s when they had been extirpated from other parts of their range.

Quaggas also lived in the Succulent Karoo (Figure 2.12), which occurs to the south and west of the Nama Karoo. The closer proximity to the coast and the lower elevation of its mountainous areas results in this region having a milder climate than the Nama Karoo, and its northwestern extension along the Atlantic coast benefits from the cold, northbound waters of the Benguela Current, causing fogs that supplement the meager rainfall. Most locations in the Succulent Karoo usually receive an annual rainfall of less than 300 mm (12 inches); however, this rainfall is predictable from year-to-year, unlike the precipitation in the Nama Karoo. These conditions favor the success of succulents and geophytes; the latter are perennial plants that grow from bulbs and can survive droughts by storing resources in their tissues. The other plants that thrive in the Succulent Karoo are shrubs, small trees, and annuals whose seeds germinate after rain into plants that rapidly grow to maturity and produce seeds while water is still available. Collectively, these plants give the Succulent Karoo a remarkably diverse flora: approximately 5,000 species are present, many of which are endemic – occurring nowhere else.[84] Quaggas and antelopes were present in this region in the nineteenth century, and the contemporary presence of mountain zebras is a further sign that zebras can live off plants of the Succulent Karoo.[85] Grasses, though comparatively rare in the Succulent Karoo, would have been readily eaten by quaggas, and they probably also ate annual plants, the fleshy leaves and stems of succulents, and the leaves of geophytes. However, water supplies would not always have been adequate.

To the south and west of the Succulent Karoo is the Fynbos Biome, one of South Africa's outstanding natural resources, which includes both fynbos and renosterveld. The former forms approximately two-thirds of the Fynbos Biome and comprises small trees and low-growing plants, including shrubs such as ericas (heaths), proteas, geophytes, and the grass-like restioids. These distinctive plants number several thousand species – many of which occur nowhere else on Earth. Fynbos plants grow mainly in nutrient-poor soils derived from sandstone and limestone and, in turn, their vegetation provides inadequate nutrition for large herbivores: a study of climate and vegetation changes over the past 18,000 years showed that fynbos vegetation could not support large numbers of mountain zebras, and this was probably true for quaggas, too.[86] Quaggas were mainly absent from some coastal locations in the Cape Colony, where the predominant vegetation is fynbos (Figure 2.12).

The renosterveld component of the Fynbos Biome comprises grasses, geophytes, shrubs, and small trees – though none growing much taller than six feet.[87] The soil of the renosterveld, derived from shale, is fertile and would have provided nutritious vegetation for large herbivores, including quaggas and mountain zebras.[88] Currently, Quagga Project zebras range on renosterveld at Bontebok Ridge and Elandsberg Reserve.[89]

There are a few records of quaggas in the Savanna where presumably grass was the main vegetation eaten. Most of this biome occurred in the north of their range but there were patches adjacent to the Albany Thicket at the southeastern edge of the quagga's range.[90]

Quaggas were present south of the Karoo and north of Port Elizabeth in the Zuurberg and Albany Thickets.[91] Vegetation in the Zuurberg comprises forests, thickets, fynbos, and grassland, but one of the common grasses was said to be "sour" (hence the "Zuur" in Zuurberg).[92] In the nineteenth century, zuurveld grass was described as "coarse and innutritious" for horses, which suggests that it may not have been a staple for quaggas.[93] Nonetheless, the Zuurberg area of the Addo Elephant National Park has supported an introduced population of mountain zebras on its diverse vegetation, which adds credence to Pringle's suggestion that the scarcity of large animals in the Zuurberg in 1821 was due "much more to the incessant pursuit of the huntsman than their aversion to the coarse herbage."[94] The Albany Thicket biome extends inward from the southern coast and consists of low-growing trees, shrubs, vines, succulents, geophytes and grasses.[95] Plains zebras living in the Albany

Thicket preferentially ate grass, but consumed other plants as well, and undoubtedly hungry quaggas in this biome did likewise.[96] The Albany District, like the Karoo to the north, encompasses highlands and valleys. The 1820 Settlers from Britain colonized this region, a group that included the poet Thomas Pringle. Arriving on June 29, 1820, at what he described as the "Promised Land," Pringle contrasted the snow-capped mountains, whose lower slopes were sparsely covered with bushes and grasses, with the verdant valley below:

But the bottom of the valley, through which the infant river meandered, presented a warm, pleasant, and secluded aspect; spreading itself into verdant meadows, sheltered and embellished, without being encumbered, with groves of mimosa trees, among which we observed in the distance herds of wild animals – antelopes and quaggas – pasturing in undisturbed quietude.[97]

Not all environments were so favorable: some locations where quaggas lived currently experience cold and drought, and there were past times when both these conditions were even more pronounced.

The Paleo-Agulhas Plain south of the current South African coastline was almost certainly a habitat for quaggas, gnus, and other antelopes whose bones have been found at coastal locations. This extensive low-lying region existed when much of the Earth's water was sequestered in vast ice sheets. Before its submergence, the Paleo-Agulhas Plain would have provided grassland for quaggas and a means for their migration east and west.[98]

Populations of quaggas would have varied depending on environmental conditions: one estimate was that they might have numbered hundreds of thousands in the eighteenth century.[99] There is no evidence for this estimate, but if quaggas had occurred at the population densities recorded for other plains zebras (just less than one to just over twenty plains zebras per square kilometer) then there could have been hundreds of thousands over their range (approximately 576,000 square kilometers, shown in Figure 0.2).[100]

There must have been times of environmental stress which reduced the numbers of quaggas: fossils from Boomplaas Cave in the Western Cape Province show a decrease in grassland in that area at the beginning of the Holocene (about 11,000 to 12,000 years ago) that was so severe that some species of grazing animals went extinct, and the populations of mountain zebras and quaggas were greatly reduced.[101] Genomic evidence also points to a reduction in population size ending about

10,000 years ago.[102] These changes occurred over the long-term, but even on a shorter time scale, periods that provided water and grazing would have alternated with droughts when herds of quaggas moved over long distances in search of sustenance, only to perish if they found insufficient resources.

The Lives of Quaggas

Early observers provided little information about the social lives and foraging habits of quaggas in their natural settings, but it is possible to extrapolate what their existence might have been like by considering this sparse record alongside what is known about other plains zebras.

Herds of quaggas often numbered twenty to a hundred, although even larger groups existed.[103] Such herds would have comprised several breeding groups, and each breeding group probably had the same composition as those of other plains zebras: a stallion, several mares, and their foals. A moving herd, probably led by a mare, walked in a "single-file march" which gave way to a "squadron-like wheel of the troop when disturbed."[104] As noted in Chapter 1, plains zebras travel long distances in search of food and water, and this was the case for quaggas, too, which in Harris's account were collectively seeking new forage:

Moving slowly across the profile of the ocean-like horizon, uttering a shrill barking neigh, of which its name forms a correct imitation, long files of Quaggas continually remind the early traveller of a rival caravan on its march ... Bands of many hundreds are thus frequently seen during their migration from the dreary and desolate plains of some portion of the interior which has formed their secluded abode, seeking for those more luxuriant pastures, where, during the summer months, various herbs thrust forth their leaves and flowers, to form a green carpet, spangled with hues the most brilliant and diversified.[105]

The "shrill barking neigh" in Harris's description can be compared with Gordon's observations: "their sound is a barking yelp, quickly succeeding each other, with many kwak, kwak, noises."[106] The foal that Gordon illustrated (Figure 2.3) had "a yelp similar to that of a jackal," and an old quagga made a sound "which resembled the barking in the distance of a little puppy."[107] Describing the quagga mare at Kew Palace, Edwards noted, "The noise it made was much different from that of an ass, resembling more the confused barking of a mastiff-dog."[108] Pringle wrote, "The cry of the *Quagga* (pronounced *quagha* or *quacha*) is very different from that of either the horse or ass," and in his poem "Afar in the Desert," he described it as a "shrill whistling neigh."[109] Many videos on the Internet show that other plains zebras also vocalize with barking sounds.

At the eastern end of their range, between the Orange and Vaal rivers, quaggas occurred together with Burchell's subspecies, as seen by the explorer Thomas Baines (1820–75) in 1850: "The plains, about two miles from the [Vaal] river, presented a scene that must be witnessed to be conceived: innumerable herds of quaggas, both of the bonte [striped] and half-striped varieties, stood quietly gazing as we passed or cantered leisurely out of our way, sometimes crossing the road within a hundred yards of us."[110]

Referring to "the Quagga" in the singular, as many authors did, Harris recorded that, although quaggas did not associate with other zebras, they kept company with wildebeests and ostriches. "It occurs in interminable herds; and although never intermixing with its own more elegant congeners [plains zebras], is almost invariably to be found ranging with the white-tailed Gnoo, and with the Ostrich, for the society of which bird especially, it evinces the most singular predilection."[111]

Harris was intrigued by the ways these animals moved together:

At every step incredible herds of Bontebucks, Blesbucks, and Springbucks, with troops of Gnoos, and squadrons of the common, or stripeless, Quagga, were performing their complicated evolutions; and not unfrequently, a knot of Ostriches decked in their white plumes played the part of a general officer and staff with such strict propriety, as still further to remind the spectator of a cavalry review.[112]

Quaggas and ostriches also associated, even when gnus were not present.[113] Lichtenstein described a flock of about thirty ostriches together with about eighty to one hundred quaggas:

As we approached them we were seen by the ostriches, who immediately took to flight, and were followed instinctively by the quaggas; for how different soever these animals are in their habits, they have a great attachment to each other, and are almost always found together. The quaggas follow the ostriches, as I have already mentioned, because the latter can see to so great a distance, and therefore sooner discern food or danger; the ostrich, on the other hand, likes to associate with the quagga, because his dung attracts a sort of large beetles, which are, to this bird's palate, great dainties.[114]

Quaggas providing food for dung beetles eaten by ostriches was not the only complementary aspect associated with feeding: plains zebras eat tall, coarse vegetation and leave the more tender plants for gnus, and this practice was probably true for quaggas, too.[115]

Aside from food, this association of animals might also have provided protection: Reginald Pocock suggested that ostriches could see further to detect food and predators, and that quaggas were better able to scent a stalking lion, "Hence the sensory imperfections of one species will be made good by the proficiencies of the others; and each will be benefited by the association."[116] No evidence was cited for these speculations and Lichtenstein, who had observed quaggas and ostriches in their natural habitats, was of the opinion that ostriches had the key role in protection: they were watchful animals that could see for long distances and readily took flight. Consequently, "quaggas generally attach themselves, as it were instinctively, to a troop of ostriches, and fly with them without the least idea that they are followed."[117] Lichtenstein wrote that Khoe-San hunters knew that ostriches served as lookouts for other animals: "They have a particular hatred to the ostrich, on account of his being so long-sighted and so suspicious, that he often betrays them to the antelopes and quaggas, whom they are lurking after, when they are got pretty near to them."[118] Lichtenstein further reported that when Khoe-San found ostrich nests they destroyed any eggs not taken as food – presumably to reduce future sentinel activity.[119]

The role of ostriches in detecting predators remains uncertain and quaggas, with their wide arc of vision and large moveable ears, might have been able to detect predators without additional help. Lacking evidence for the relative efficacy of vision, hearing, and smell of quaggas, ostriches, and other animals, it can at least be said that the combined senses of these species grazing together – and any birds that associated with them – would have been helpful in detecting stalking predators: all would have been alerted when an animal bolted or uttered a warning cry. Family members would probably have gathered around a mare with a young foal to protect them if a predator approached close, as occurs in other plains zebras. Quaggas were said to fight hyenas tenaciously – biting, kicking, and trampling them until they were dead.[120]

Fighting was not the only option: quaggas also fled from predators and probably could gallop at the fifty-five kilometers (thirty-four miles) per hour recorded for other zebras.[121] Harris observed: "Judging from its low, and somewhat laboured pace, the inactive spectator would pronounce the Quagga to be a slow and heavy galloper; but it is only necessary to follow its flight a few yards on horseback, to be convinced of the rate at which it covers the ground."[122]

Sparrman noted that quaggas shared with antelopes and buffaloes, "a peculiar custom of standing still at intervals during their flight, to stare at their pursuers a little, and wait for their coming up."[123] At first sight, allowing predators to approach seems disadvantageous but quaggas may have been assessing the level of threat, as running long distances to escape predators is costly in both its use of energy and reduction in grazing time, an important consideration for animals that must spend much of their time eating.[124] Overheating from running and the accompanying loss of water from sweating are additional reasons for not running long distances away from predators.

Quaggas faced predation from humans, hyenas, lions, and leopards. Parasites such as worms, flies, ticks, and lice would have sapped their strength, and microorganisms caused diseases. These challenges could occur in combination: an animal weakened by disease and ill-nourished because drought had reduced its water and food supplies would be more likely to fall prey to a predator than a fit animal.

From Origins to Extinction

The ancestors of quaggas migrated into southern Africa and occupied several contiguous biomes. Generation after generation lived, migrated, and inter-acted with other creatures through changes in the climate that affected their environments. Existing largely isolated from other plains zebras for hundreds of thousands of years, these animals developed their characteristic morph-ology. The brown coloration that is sometimes present in other plains zebras became more pronounced, stripes were reduced on their hindquarters and were lost altogether on their legs. Their appearance led some observers to believe that they were a different species from plains zebras, and their tractability in harness caused others to see them as horse-like.

For much of their existence, the only people quaggas knew were Khoe-San and Xhosa. If the total lifespan of quaggas is viewed as a day, then most of their contact with Europeans occurred after the clock registered 11:59 pm. During that last minute, Jan van Riebeeck admired them, Boddaert and Gmelin named them, Agasse, Daniell, Maréchal and others painted them, Fritsch, Haes and York photographed them, and Edwards, Harris and others described them. Quaggas were harnessed to pull carriages, corralled with domestic animals to protect them, and hunted. Their hides were transformed into an array of items, and some were mounted in museums, which also became the repository for their bones. The one thing that human creativity could not achieve was their survival.

3 · *Coat Coloration*

The true quagga was nothing but the dullest coloured and most southerly form of Burchell's zebra.

Frederick Courteney Selous[1]

Stripes, the defining feature of zebras, are so striking that they have inspired inventive origin stories and black-and-white striped warships. Scientists have been similarly creative in speculating on their functions – often in the absence of substantial evidence. It has been variously claimed that stripes: camouflage zebras, cool them, discourage flies from landing on them, aid recognition, and promote social interaction.[2]

Functions of Stripes

One of Rudyard Kipling's *Just So Stories*, "How the Leopard Got His Spots," offers a brief explanation for zebra striping:

Zebra moved away to some little thorn-bushes where the sunlight fell all stripy, and Giraffe moved off to some tallish trees where the shadows fell all blotchy ... Leopard stared, and Ethiopian stared, but all they could see were stripy shadows and blotched shadows in the forest, but never a sign of Zebra and Giraffe. They had just walked off and hidden themselves in the shadowy forest.[3]

Stripes and spots, in Kipling's explanation, served as concealment camouflage, or crypsis, against human and non-human predators. He may have based his story on the suggestion by Alfred Russel Wallace that stripes might hide zebras "when the animal is at rest among herbage – the only time when it would need protective colouring."[4] Stephen Jay Gould famously warned twentieth-century readers to be skeptical of "just-so" explanations of evolution.[5] But, before that, Charles Darwin was unconvinced by this idea of crypsis, pointing out that zebras live in the open plains.[6] In fact, zebras occur in a variety of habitats, and so proponents of crypsis would have to explain the selective advantage of stripes in each of these habitats and under a range of lighting conditions.

Another possibility is that the stripes of fleeing zigzagging zebras might confuse predators and so make it more difficult for them to catch individual animals.[7] This "dazzle" or "motion" camouflage can influence how humans perceive speed."[8] Two experienced biologists who had frequently observed zebras in the wild have noted the effects of stripes on human perception. Jonathan Kingdon observed that the width of a zebra's vertical body stripes seemed to narrow or widen as the animal altered its position.[9] Austin Roberts commented that the lighting conditions and the position of a zebra might affect its appearance to the extent that its stripes would or would not be seen.[10] The theory that movement of a striped object might bewilder an observer inspired some Allied navies in World War I to paint their vessels with alternate light and dark stripes to deceive the enemy about the speed and direction of the ships. Although confusion of a predator by the stripes on a fleeing zebra might also seem to be a "just-so" story, computer simulations show how stripes cause humans to misjudge the movement of objects.[11] If the effects are the same on predators, catching prey will be more difficult. Much more distortion is apparent when the animals cluster: three zebras moving close together produce an image that might perplex a predator chasing a fleeing herd.[12]

Zebras and non-striped animals living together provide a natural experiment to test theories about the benefit of stripes. Wildebeest and zebras frequently graze together and are hunted by lions, but whereas one study showed that both were equally likely to become prey, another showed that plains zebras have a survival advantage over wildebeests.[13] Even if stripes do not benefit zebras by confusing lions, those zebras that survive probably owe something to their other attributes: senses to detect intruders, an anatomy evolved for speed, a social organization that can help protect the most vulnerable, and the ability to kick and bite when attacked.[14]

One investigation centered on whether stripes create small-scale air currents next to the body that aid thermoregulation.[15] Sweat evaporation cools zebras. Under windy conditions, airflow over the surface of their bodies causes evaporative cooling but in the absence of wind the alternation of black and white stripes might cause small-scale air currents that would aid in evaporation.[16] During the hottest hours of the day, black stripes were 10°C to 15°C (18°F to 27°F) hotter than white stripes, and this difference could cause air movements over the body surface that dissipated the sweat and cooled the animal.[17] These differences in stripe temperature occurred near the equator in plains zebras that had well-defined stripes present over their whole bodies, and would be reduced if the striping were not so pronounced and the solar radiation were less intense – which is exactly the situation for plains zebras living near the southern end of their range. However, any effects of stripes on thermoregulation might be minor compared with cooling produced by body movement or wind.

The most compelling experimental evidence shows that stripes protect against flies. Flies are often a problem for equines: their presence in large numbers can be so annoying that they distract animals from grazing. Bites from flies are not only painful and result in loss of blood but can transmit several deadly diseases including African horse sickness and trypanosomiasis. Tsetse flies, the vectors of trypanosomes, are less attracted to surfaces with black and white stripes than they are to either black or white surfaces; this observation suggests a major selective advantage for zebras' stripes in habitats where tsetse flies are common.[18]

Horseflies (tabanids) are also troublesome to zebras and here, too, tests of the attractiveness of black and white striped patterns to horseflies showed that the narrower the stripes, the less attractive the patterned surfaces were. The width of the narrow, uninviting stripes corresponded to those on zebras, suggesting that the coat pattern of zebras evolved to make them less attractive to horseflies and so lessen fly bites and the likelihood of contracting fly-borne diseases.[19] Possibly, an insect's own movement as it flies towards a zebra would lead to the same confusion that stripes cause humans in computer simulations.[20]

Further support for the protective role of stripes against biting insects came from investigations that correlated the stripes in equines with the presence of biting insects and that specifically rejected any role for stripes in camouflage, thermoregulation, sociability, and the avoidance of vertebrate predators.[21] This experimental evidence is supported by the observation that the subspecies of plains zebras, native to equatorial regions where flies are abundant, have well-defined stripes over most of their bodies, whereas populations of plains zebras that evolved in the cooler and drier conditions of southern Africa where flies are scarcer have fewer and less-defined stripes.

Partially striped zebras selectively bred by the Quagga Project could be used to determine whether flies are less attracted to black and white stripes than they are to a uniform color by analyzing the attractiveness to biting flies of the striped forebodies compared with the unstriped hind bodies of these animals. Similarly, the presence and absence of stripes on the same animal might test the effects of stripes on the creation of air currents and thermoregulation. Naturally, a control for these experiments with partially-striped zebras would be parallel observations on fully-striped zebras kept in the same environment.

The evidence is strong that stripes deter biting flies, but might they also enable zebras to recognize one another? Each zebra has a unique pattern of stripes that enables people to identify individual animals within a herd, and presumably zebras can do likewise. But, in the absence of stripes, wild horses can also recognize members of their herd.[22] Although this observation suggests that stripes may not be essential for recognition by zebras in most instances, foals that have lost contact with their mothers appear to identify them by sight, rather than smell.[23] Alfred Russel Wallace similarly suggested that stripes, "*may* be of use by enabling stragglers to distinguish their fellows at a distance."[24] This speculation overlooks real-life conditions where zebras stay together in a breeding group or herd: being a straggler at a distance is dangerous where predators are present.

Stripes may, however, be important in the social structure of herds in ways other than recognition. Mares groom their foals and later these foals will groom other members of their breeding group – creating a social environment. Grooming often occurs at the junction between the neck and shoulder, which suggested to Kingdon that zebra's stripes originated here to show a preferred location for this interaction, but later spread to other parts of the body to make an even more powerful statement of sociability.[25] Kingdon viewed the social behavior of zebras as being important for them and conjectured that stripes are the key to this behavior, initially fostering physical contact between mother and foal but then assuming a broader significance by promoting bonding between many zebras.

Developmental Origins of Stripes

Zebra stripes are formed from alternating bands of dark and light hairs that cover their skins. The same banding occurs in the manes, which are composed of bristly hairs that stand upright from the necks of the animals, unlike the manes of horses, which lie flat. Zebra stripes are generated between the third and fifth weeks of embryonic development, but it is only at about the eighth month that the dark pigment, melanin, is deposited in the hairs destined to form the dark stripes; hairs of the white stripes lack this pigment.[26] The number of stripes and their widths vary between the three species of zebras; for example, Grévy's zebras typically have more than twice the number of stripes present in plains zebras, and the individual stripes are narrower (Figure 1.3).

Jonathan Bard investigated the factors determining the numbers of body stripes by examining their formation during embryonic development.[27] He found that stripes about 0.4 mm wide, corresponding to a thickness of about twenty cells, formed in all three species of zebras, but that stripe formation occurred at different times in the three species. In Grévy's zebras, stripes form in the fifth week: at this time a stripe repeat of 0.4 mm along the length of a 32 mm long embryo yields about eighty body stripes –resulting in the many narrow stripes of this species. In contrast, the stripe pattern of plains zebras was initiated when the animal was about three weeks old: in this case, a stripe repeat of about 0.4 mm along the length of the embryo (11 mm) yielded between twenty-five and thirty body stripes. Mountain zebras were intermediate between the other two zebras: 0.4 mm wide stripes formed along a 3.5-week-old embryo of length 14–19 mm caused about forty-three body stripes to form.

The striping pattern on the heads and necks of quaggas corresponds to that in plains zebras, and so Bard viewed quaggas as being plains zebras in which stripes did not form at some locations. Other variations in coat coloration between quaggas and heavily-striped plains zebras were that in quaggas white stripes were replaced with reddish-brown stripes, and that the legs and bellies of quaggas lacked pigments altogether, except for the ventral line.

Aside from differences in striping, individual zebras may exhibit major alterations in coat coloration; for example, there are albino plains zebras in which dark pigmentation is reduced and melanistic plains zebras which have a superabundance of dark pigment in their hairs. There are degrees of pigmentation in both these conditions, so that some albinos are entirely white while others have light-gray stripes in place of black stripes; likewise, some melanistic plains zebras have more overproduction of pigment than others.[28]

In rare instances, spotted zebras occur: a black zebra with white spots and a zebra with both stripes and spots.[29] Describing a plains zebra that was black with white spots arranged in rows corresponding to the positions of white stripes, Bard observed that this pattern could only occur if there had been a faulty development of white stripes on a dark background and concluded from this arrangement that zebras are black animals with white stripes.[30] Quaggas provide further evidence for this viewpoint, as the unstriped hindquarters can be considered to be an example of the dark background.[31] Others have similarly viewed this subspecies as dark animals with pale stripes, and certainly the photograph of the mare in London Zoo (Figure 5.1) gives this impression.[32] Some authorities, however, regard zebras (including quaggas) as having dark stripes against a light background.[33] One can sidestep this issue by describing zebras as having black and white stripes, or, in the case of quaggas, dark brown and lighter brown stripes together with unstriped bellies and legs.

Stripe Variation in Plains Zebras

The stripes of mountain zebras and Grévy's zebras are black and white over much of the animal, but plains zebras are more variable. In equatorial Africa, plains zebras have starkly black and white stripes over the whole animal, including the legs. Further south, brown shadow stripes are present within the white stripes and there is a progressive reduction in striping. This trend continues into southern Africa, where Burchell's subspecies have legs with fewer stripes, dark stripes on their bodies that are fewer and browner, and shadow stripes that are more prominent. Biologists term this sort of gradation across the range of a species a "cline." Some observers of this cline viewed quaggas not as a separate species but as the most southern population of plains zebras – being the least striped and most brown subspecies.[34]

Larison et al. correlated variations in coat coloration of plains zebras with temperature; higher temperatures are associated with well-defined black and white stripes over the body and the entire length of the legs (Figure 3.1). They noted that there is diminishing intensity of striping in plains zebras from the north to the south of their range, accompanied by a reduction in number of stripes.[35] A broader study of stripes in the seven extant equines and their subspecies (the three species of zebras, African wild asses, and, in Asia, kiangs, onagers, and Przewalski's horses) concluded, "Striping was associated with six or more consecutive months that fell within a temperature range of 15°C to 30°C and humidity between 30% and 85%."[36] These conditions favor the breeding of many species of flies and argue that stripes evolved to counter biting flies.[37]

This variation in coat coloration leads to the question: what caused quaggas to lose stripes and evolve a chestnut color on their upper bodies? The earliest "just-so" explanations about quaggas involved camouflage.

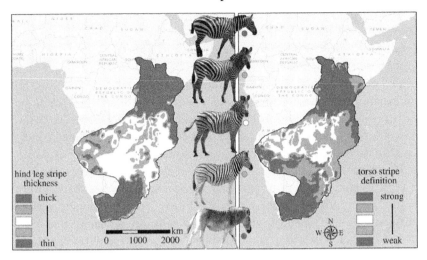

Figure 3.1. "Predicted levels of hind leg stripe thickness (left) and torso stripe definition (right)." From figure 2 in Larison et al, "How the zebra got its stripes: a problem with too many solutions," Royal Society Open Science January 1, 2015. Attribution-Share Alike 4.0 International license. https://doi.org/10.1098/rsos .140452. (A black and white version of this figure will appear in some formats. For the color version, please refer to the plate section.)

Pocock considered that the reduction of stripes on their bodies and the absence of stripes on their bellies and legs would have concealed quaggas in open country.[38] Rau also speculated that the coat coloration of quaggas gave them a selective advantage in open country.[39] These suggestions were unaccompanied by supporting evidence.

A better explanation begins with flies. Horseflies and tsetse flies require warmth and humidity for much of the year, and in the tropical regions where these conditions occur, plains zebras have well-defined stripes over most of their bodies. Burchell's subspecies, living in the cooler, drier conditions in southern Africa where there are fewer horseflies and tsetse flies, have fewer stripes and these are less pronounced than in plains zebras living in tropical regions. Quaggas represented a further continuation of the trend in reduced striping and browner coloration. Horseflies often land on the legs of zebras and bite there but can be deterred by stripes. With a reduction in horseflies, there would be less selective advantage to having leg stripes, and it is notable that Burchell's subspecies have few leg stripes and quaggas had none. These correlations and the experimental evidence cited earlier for the effects of zebras' stripes on flies seem compelling.

Not only did quaggas live at the southern end of the plains zebra's range, many of them lived in the elevated lands of the Karoo and the Orange Free State. Today, these areas are often dry and cold, conditions that were exacerbated during the last two Ice Ages when changes in climate that caused glaciers to form in colder regions of the world had major effects on temperatures and aridity in southern Africa.[40] Genomic studies indicate that quaggas diverged as a separate subspecies over the past 120,000 to 356,000 years, and so it is reasonable to think that environmental conditions could have affected their evolution during the last two Ice Ages.[41]

During the last Ice Age temperatures in the Karoo were as much as 5°C (9°F) colder than at present and precipitation was less than current levels.[42] The other habitats of quaggas might have experienced similar changes. Droughts might have isolated a population of plains zebras from others of their kind in a colder habitat where conspicuous black and white stripes had less selective advantage as there were fewer flies. In the absence of breeding with populations of plains zebras that were more striped, quagga coat coloration would have become established.

Another possibility is that droughts that depleted water supplies and greatly reduced grazing land might have killed off a large percentage of the geographically isolated population – allowing particular alleles for the chestnut coat color and reduced striping to become overrepresented by chance and not by selection. This process of genetic drift (the neutral theory of evolution) occurs in small populations and, like selection, drives evolutionary change.[43] These two possibilities are not mutually exclusive: as others have noted, the coat coloration of quaggas may have arisen from an interplay between genetic drift because of geographical isolation and selective responses to the environment.[44]

In Rudyard Kipling's origin story, zebra stripes were a response to the environment: "after another long time, what with standing half in the shade and half out of it, and what with the slippery-slidy shadows of the trees falling on them, the Giraffe grew blotchy, and the Zebra grew stripy."[45] Kipling can be parodied to provide a more current explanation of coat coloration in quaggas: "after many millennia, what with living as an isolated population subject to selection and genetic drift, and what with the cold, dry conditions that kept flies away, the Quagga grew less stripy."

Loss of Stripes

The last word about quagga coloration goes to Darwin, who considered the question: are zebras striped horses, or are horses zebras that have lost their stripes? Darwin had no doubt about the answer, although he posited loss – not gain – of characteristics over the course of evolution. On equines in general he said: "For myself, I venture confidently to look back thousands on thousands of generations, and I see an animal striped like a zebra, but perhaps otherwise very differently constructed, the common parent of our domestic horse, whether or not it be descended from one or more wild stocks, of the ass, the hemionus, quagga, and zebra."[46]

Darwin's belief that horses, like quaggas, had lost the striping of their equine ancestors turns around the question about whether quaggas were more horse-like or donkey-like. In having lost their stripes, both domesticated equines were actually quagga-like.

4 · Quaggas, Zebras, and Humans in Southern Africa

A moth came to our hut's fire, our mothers exclaimed, they told us about it ... [moths] which dwell with the game, they are those which are wont to come to our fire ... they come to burn in our fire; while they feel, that, the game with which they live, its flesh will roast at our fire. Therefore, they come to tell us about it, that, we shall put the flesh of the game to roast at the fire ...

Dai!kwain[1]

The indigenous people of southern Africa had no domesticated equines. To them, quaggas were wild animals and among those they hunted. Some of their stories tell of hunting and – as in Dai!kwain's words – the anticipation of feasting. In Europe, however, horses and donkeys were critical for war and transportation. Seeing zebras, including quaggas, in southern Africa, Europeans wondered: could these eye-catching equines be saddled and harnessed? Whether zebras, including quaggas, could be domesticated remains an open question. Despite a few attempts, they remained wild, and their most significant relationship with people was as prey.

Quaggas' Human Neighbors

The first recorded people in southern Africa were the hunter-gatherers now collectively known as Bushmen, who lived throughout the subcontinent, including the dry interior where quaggas roamed. They were able to survive in arid areas by harvesting wild plants and catching animals, whose likenesses they depicted in rock paintings and engravings (Figure 2.2). These hunter-gatherers included the |Xam of the Karoo, whose stories about quaggas are recounted in Chapter 7. Quaggas were also known to the Khoekhoe, who herded cattle, sheep, and goats, but did not cultivate. They probably had origins among foragers in what is now Botswana and spread throughout the subcontinent seeking new pastures for their herds. They lived in the semi-arid interior and in the coastal areas on the western side of the subcontinent where rain fell in the wintertime.[2] Other people, who spoke Bantu languages, used iron, cultivated crops, and lived in settled villages, took up residence in south-east Africa around the beginning of the Common Era. These were the ancestors of the people who identify today as Zulu, Xhosa, Swazi, Tsonga, Tswana, North and South Sotho, and Venda. They herded cattle, sheep, and goats, but also cultivated. Their crops, which had been domesticated in tropical Africa, required summer rainfall, so they settled on the eastern side of the subcontinent, first establishing themselves in small bands and later merging into chiefdoms. Concentrated in the east of the arid territory where quaggas lived, the agro-pastoralists had less interaction with them, nonetheless, Xhosa gave a name to quaggas, "iqwarha."

The Portuguese explorations of the fifteenth century introduced the final human actors in the story of quaggas. In February 1488, the Portuguese mariner Bartolomeu Dias (1450–1500) rounded the south-west corner of Africa that he called the "Cape of Storms." Later, mariners and mapmakers renamed the Cape more optimistically "Good Hope." After Dias, in 1498, his compatriot Vasco da Gama was the first European to sail into the Indian Ocean, and others followed. Soon, Europeans voyaged regularly to the Indian Ocean and the Pacific coasts of Asia, and the Cape of Good Hope became a necessary victualing stop.[3]

Europeans took the water as free and in principle offered goods in exchange for Khoekhoe stock, but, too often, they used guns to get what they wanted. The Dutch East India Company, profiting richly from its trade in Asian markets, founded a "refreshment station" at the Cape. The Dutch outpost at the Cape was not initially meant to be a colony, and Europeans living in the castle in what is now Cape Town would have had little contact with quaggas in the interior. The European presence, however, could not be confined to the castle. Finding it difficult to provision ships through company efforts alone, in 1658 the Dutch East India Company allowed "free burghers" to take up independent farming on fertile lands near the castle. These settlers steadily expanded into Bushmen hunting grounds, Khoekhoe pastures, and eventually the fields and farms of Bantu-speakers.[4]

Within a few decades, this frontier of white settler expansion under-mined Khoekhoe herding societies. Slavery was legal in South Africa and exploited people were imported from all around the Indian Ocean world. Most of the people rich enough to buy chattel slaves were in cities or on the large exporting farms in the arable district near Cape Town. The indigenous people were not legally enslaved or sold as chattel property. But, on the frontier, the descendants of the defeated Khoekhoe and Bushmen became dependent, largely unfree, laborers for white farmers.[5]

The chief institution of European frontier farmers was the "commando," the citizens' militias that raided the Africans' herds, hunted wild animals, and fought against any groups that opposed their expansion. In the eighteenth century, the foraging Bushmen, who hunted with impunity in the farmers' herds and made it impossible for settlers to live securely on the far frontier, became the target of repeated commando actions. Settlers, with the support of the Dutch East India Company, promulgated genocide against Bushmen bands. As they hunted down adults, they took children to serve as laborers.[6] Modern South Africa has its origins in the early Cape Colony and its subjugation of the indigenous people. Colonial history bequeathed not only the racial order but also the market economy and the transformation of nature.

Africans Hunting Quaggas

Remains of zebras have been excavated from caves occupied by humans over many millennia. Depending on location, these include bones of mountain zebras and plains zebras (including quaggas). Teeth of plains zebras, quaggas, and giant Cape zebras (a large extinct subspecies of the plains zebra) all occur in the Wonderwerk Cave in the Kuruman Hills in the Northern Cape Province, where the evidence of fires over the past 800,000 years argues for a long occupation of this site by humans.[7] Zebra teeth that are at least 20,000 years old – probably from both mountain zebras and quaggas – have been found in the Boomplaas Cave of the Western Cape.[8]

Khoekhoe and Bushmen probably used similar techniques when hunting quaggas, but the account that follows describes the weapons and practices used by the Bushmen. Men hunted quaggas and other animals using bows and arrows that Lichtenstein described in detail.[9] Bushmen fashioned a bow from a piece of hard wood about five-foot-long, tapered it towards its ends, and strung it with twisted intestines taken from hunted animals. They used a strong reed to make an arrow, approximately three and a half feet long, which was notched at its base to accommodate the bowstring. Entrails wrapped around the reed just above the notch strengthened the shaft, and a feather from a bird of prey was attached to the base to stabilize the arrow's flight.

Each arrow was tipped with, "a hard hollow piece of bone, commonly the thigh bone of the antelope, sharpened to a point, or a small triangular plate of iron."[10] Bushmen made a poison from snake venom, sap of *Euphorbia* plants and an extract of *Haemanthus* bulbs. They stirred these components with a stick in a heated hollowed-out stone and rubbed the mixture on to the arrow tip where it dried to a hard substance.

Two modifications ensured that the poisoned tip stayed in the body when the arrow hit the animal. A cut was made in the shaft just below the tip so it would break at this part – leaving the tip in place if the arrow hit a bone and bounced back from it. Close to the cut on the shaft was a hook fashioned from a quill that would hold the tip if the arrow struck a softer part of the animal's body. The poison produced tissue swelling at the wound site, which also prevented the tip from leaving the body while the poison did its work. Bushmen marked their arrowheads so they could identify which hunter had killed an animal.[11] Hunters carried arrows in a quiver made from the hollow stem of an *Aloe* plant that was covered, whole or part, in leather (Figures 4.1 and 4.2). Extra arrows were carried around their heads (Figure 4.1).[12] Attached to a leather strap, the quiver was carried over the left shoulder – providing easy access to the arrows so that a hunter could shoot five or six during a minute, holding the bow with his left hand and drawing back the bowstring and arrow with his right.[13]

Petrus Borcherds and Anders Sparrman gave briefer descriptions, citing different dimensions. Sparrman described bows as being approximately three feet long and arrows being about one and a half feet long.[14] Borcherds recorded that there were two lengths of bows: approximately two feet and four feet long with strings made of tendons; he described arrows as being about one-and-a-half to two feet long whose points were a piece of iron or a sharpened ostrich bone that had been covered with poison made from snake venom or plants.[15] Sparrman depicted a range of weapons used by the Khoe-San (Figure 4.2).[16] Spears were seven to eight feet long and were used in game drives and to force lions to abandon the carcasses of their prey; they might also have been used to dispatch animals that had been sickened with poison but had not yet died.[17] Some hunters also carried a kierie (club) that they threw to kill small game.[18]

Careful stalking of the prey was important; hunters operating individually or in small numbers could approach to within fifty to sixty paces of animals "by almost crawling along the earth upon their bellies,

Figure 4.1. Bush-men [Khoekhoe] armed for an expedition. Plate 2 in Daniell, African Scenery and Animals, 1804–1805. Courtesy of the British Museum. (A black and white version of this figure will appear in some formats. For the color version, please refer to the plate section.)

strewing their bodies and garments over with dust, that the colour may not betray them, and never moving if they see that the animal appears to be looking that way."[19] Bushmen could shoot arrows up to 200 paces and could hit their target "with a tolerable degree of certainty, at the distance of fifty, or even a hundred paces."[20]

Bushmen also stampeded animals – including quaggas – to a location where fences forced them into pitfalls.[21] These game drives involved many people, sometimes including women and children.[22] Hunters occasionally used a different technique for antelopes, particularly springboks, and it might have been used for quaggas, too: a group of several hundred hunters formed an extended circle around animals on the veld. As the hunters constricted the circle, they made an opening through which the antelopes stampeded but, in so doing, impeded each other, enabling hunters to spear them.[23] Sometimes when animals were stampeded, long sticks with ostrich feathers attached to them directed the flight of the game, which presumably mistook these objects for humans.[24]

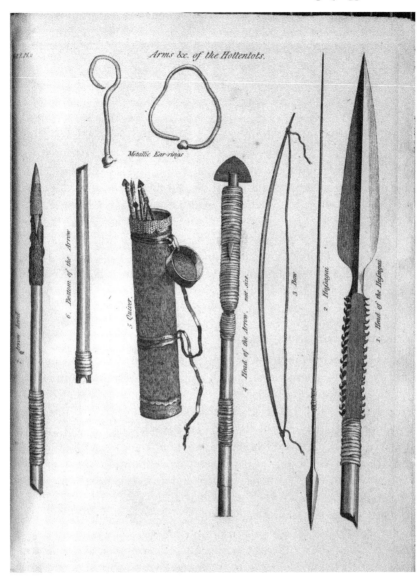

Figure 4.2. Arms &c. of the [Khoekhoe]. Plate 2 in Sparrman, A Voyage, Volume 1.

Bushmen employed other techniques for catching wild animals, and some of these may have been used for quaggas. They dug game pits containing vertical sticks sharpened at their skyward ends to pierce animals that fell through the grass covering of the pit.[25] Hunters poisoned the water sources of animals with the sap from *Euphorbia* plants and then tracked down sickened animals whose meat was edible since the effects of the poison were confined to the intestines.[26] Sometimes other predators did the killing: Harris observed that spear-wielding Bushmen drove away lions from the quaggas they had killed and took the meat for themselves – a practice that earned the Bushmen his description of "two-legged devourers of carrion."[27]

Quaggas were hunted for their meat, which was relished by the Khoe-San.[28] Indigenous people used dried tendons to make bowstrings, sharpened bones into arrow tips, used entire hides as karosses (cloaks) and fashioned pieces of hide into shoes. Sparrman enthused, ". . . these field shoes, as they are called, [are] made of almost raw leather," and he noted, "They fit as neat upon the foot as a stocking and at the same preserve their form." Sparrman wore these shoes and brought them back to Sweden as a model explaining that, "Whatever is useful, whether it comes from *Paris* or the country of the [*Khoekhoe*] alike deserves our attention and imitation."[29]

Colonists Hunting Quaggas

Zebras were frequently hunted. Hunters with rifles – often on horseback and so able to outrun their prey – shot them for hides, meat, and "sport."[30] Guns were widely held and used in the Cape Colony.[31] Initially, they were restricted to soldiers and employees of the Dutch East India Company, but by the end of the late seventeenth century they had spread to settlers and Khoekhoe.[32] The first soldiers at the Cape had matchlock muskets, but these heavy and unreliable weapons were replaced by Dutch and later French and British flintlock muskets.[33] The most successful of these weapons was the "Brown Bess" flintlock musket which was introduced when the British occupied the Cape Colony in 1795.[34] These long and heavy muzzle-loading firearms, often termed "roer" in southern Africa, remained important weapons there until breech-loading rifles using cartridges were adopted in the mid-nineteenth century. Consequently, it was muzzle-loading guns firing lead balls that wrought such a devastating effect on the wild animals of the Cape Colony, even before the introduction of breech-loading rifles.[35]

The maintenance and use of a muzzle-loader required that the hunter carry a loading rod and several pouches of supplies.[36] Loading these smooth-bore guns was a skillful process. Hunters bit the end off a paper

Figure 4.3. Boors returning from hunting. Plate 11 in Daniell, African Scenery and Animals, 1804–1805. Courtesy of the British Museum. (A black and white version of this figure will appear in some formats. For the color version, please refer to the plate section.)

cartridge containing gunpowder; they put a small amount of powder in the gun's pan and poured the remainder down the barrel. Next, they inserted into the barrel a spherical lead ball wrapped in cloth or leather together with the paper wrapping of the cartridge which ensured that the ball did not roll out of the barrel if it was tipped downwards. Finally, hunters thrust the whole down the barrel with a loading rod.[37] This lengthy procedure made it important that a first shot killed an animal outright or wounded it sufficiently that a second shot could be fired before it escaped.

Some hunters loaded and shot from horseback – a feat that required considerable skill. "In chasing the different kinds of game, he rides in full gallop till within gunshot, when, throwing the bridle upon the mane, the horse at once stops short, stands firm, and he fires from his back."[38] Hunters rode mostly on Cape horses that in size (averaging a shoulder height of about 57 inches) and brown color were just a little larger and not too different in appearance from the quaggas they pursued and outran.[39]

Double-barreled guns were introduced into southern Africa in the 1820s and had a longer range than Brown Bess.[40] Some of these guns had one smooth barrel to fire buckshot, and a rifled barrel that caused the bullet to spin in flight, improving accuracy and effective range. Cornwallis Harris hunted with one of these guns.

Breech-loading rifles introduced in the 1860s were superior in range and accuracy to Brown Bess, yet muzzle-loading guns used powder and lead that were readily available, could be repaired by their owners, and were reliable. By contrast, breech-loading rifles were more expensive, used specialized cartridges, and had to be repaired by gunsmiths.[41] Tellingly, Selous, who began his African hunting career in the 1870s when breech-loading rifles were available, chose to use a large-bore, muzzle-loading gun.[42] Other hunters made a similar choice: 98 percent of the ammunition sold during 1873 in Hopetown, now in the Northern Cape Province, would have been used in muzzle-loading firearms.[43] Hopetown in 1873 was not far in time and distance from the last of the quaggas, and so these animals must have fallen to the same old-fashioned gun technology that had earlier extirpated their kind from the south of the Cape Colony.

The frontier farmers who usually preferred mutton and beef – or, if eating game, antelope – reportedly disdained quagga and zebra meat.[44] This viewpoint probably reflected the cultural aversion of many Europeans to eating horsemeat. Aside from the taboo against zebra flesh, the meat often had a thick layer of yellow fat so unappealing that people unaccustomed to it sometimes vomited.[45] European travelers who experimented with eating quagga and zebra meat recorded different opinions. Some described quagga meat as having a sweet taste.[46] Gordon proclaimed that roasted mountain zebra meat had a "very pleasant" taste, but Harris compared quagga meat unfavorably to horse meat, "its flesh, although infinitely more yellow, rank, and oily than that of a horse, was greatly esteemed by the [Khoekhoe]."[47] However, a contemporary of Harris recorded a preference for grilled quagga over horsemeat, and possibly the broiled slice of quagga's tongue that was described as "sole leather" might just have been overcooked.[48]

Quagga meat did not go to waste: Gordon observed that "Our farmers do not like it, but the [Khoekhoe] do,"[49] and the explorer William Somerville (1771–1860) reported likewise, "The [Khoekhoe] regaled themselves with the flesh of the Quacha, in their estimation the most delicate of any that the country produces."[50] Quagga meat that "was so warmly praised by my [Khoekhoe], as being excellent *meat*," led Burchell to eat a quagga steak that he pronounced to be, "good and palatable. It was tender, and possessed a taste which seemed to be between that of beef and mutton."[51] Burchell, however, further commented that he could not:

... resist altogether the misleading influence of prejudice and habit; and allowed myself, merely because I viewed this meat as horseflesh, to reject food which was really good and wholesome. In this respect, the [Khoekhoe] are much wiser than the Boors, who reject it for the same reason with myself, but who, nevertheless hunt these animals for the use of their [Khoekhoe] and slaves.

In short, it appears that well-prepared quagga meat was quite tasty – as the servants and slaves who relished it knew full well. All indications are that quagga meat was a valued part of the diet of indigenous people, and probably this had been true long before colonists arrived.

Colonists made extensive use of quagga hides: initially, they were made into relatively inexpensive quotidian items such as ropes, harnesses, whips, halters, bags, thongs, clothing ,and *veldskoen* (sometimes written as *veld-schoenen* or *velschoons* – rough rawhide shoes with the hairs on the exterior and the thicker parts of the hide providing the soles),[52] or were cut into *riempies* (strips) that were used to bind things together.[53] Linking the rural and industrial worlds, quagga skins were used as connecting bands for machinery.[54] Untanned quagga hides covered unglazed windows at night on frontier stock farms.[55] Hides would have also been used as decorations and rugs or would have been sold to museums and private collections. However, quagga hides were offered for sale in an 1840 catalog price list for considerably less than the hides of mountain zebras and plains zebras.[56] Presumably, people who displayed zebra skins wanted them to be fully striped. Edwards made a similar observation, "I suppose the skins of the females [quaggas] are not counted so beautiful as those of the males [mountain zebras]."[57] A few of the hides ended up as taxidermy specimens, as was the case for the hide of a full-term quagga fetus that Sparrman took to Sweden.[58] In the mid-nineteenth century, quagga skins became valued as high-quality leather for hunting boots but that is part of the story of their extinction.

Harnessing Quaggas (and Other Zebras)

An early twentieth-century author speculated that quaggas, "could have been domesticated and made serviceable to man" and commented that: "its native home is a continent infested with the Tsetse Fly, and cursed with horse-sickness, where it might have been of immense service to those who have been the means of its extermination."[59] The possibility of domestication is the great "what-if" in quagga history. Some colonists did use them for traction but, for a combination of reasons, harnessed quaggas remained a curiosity and were never domesticated.

Once the Dutch had established themselves in Cape Town and had built a fort, they explored the hinterland. Jan van Hawarden, leading a trek in November 1658, reported encouraging Khoekhoe to capture "horses" for them; Jan van Riebeeck (1619–77), who founded the initial Dutch settlement of what became the Cape Colony, described these as "coloured and marked in an extraordinary way seen nowhere else in the world and, because of their rarity, would probably be beyond price."[60] Van Riebeeck enthusiastically described a zebra that Pieter Meerhoff saw during a later expedition,

The horse's whole body was an extraordinarily beautiful dapple grey, except that across the crupper [rump] and buttocks, and along the legs it was most beautifully and strangely streaked with white, sky-blue and a brownish-red. It had small ears just like a horse's, a fine head and slender legs like the best horse one could wish for. [61]

Van Riebeeck's wonder over these "horses" had a practical aspect: he desperately required more horses and mules, but his plea, "We are therefore still urgently in need of another 6 or 8 horses" had gone unanswered by the Dutch East India Company.[62] Nor would the Khoe-San supply him zebras as he judged that, "Apparently they are beginning to realise more and more that we would thereby be the better able to keep them in submission."[63] Eventually, Van Riebeeck's request was fulfilled and by 1670 the Cape had more than fifty horses.[64] Well supplied, therefore, with domesticated equines, interest in domesticating quaggas and zebras diminished.

City households in Cape Town and wealthy wheat and wine producers in the winter rain-fed arable district near the city were more tightly connected to the global economy and better capitalized than the stock farmers in the interior who lived at the edge of the global commercial economy. These farmers traveled to Cape Town every few years to sell hides, rhinoceros tusks, and elephant ivory for enough cash to purchase what they could not make for themselves, including guns and ammunition. Frontier farmers could use oxen to haul their families and goods on these periodic trips to Cape Town, but they needed equines for local transportation, hunting, farming, and especially for their commandos.[65]

Some frontier farmers attempted to incorporate quaggas into their livestock. To do this, they had to capture them alive, with a lasso or by luring them with hay into enclosures.[66] Possibly, too, they caught quaggas when they mingled with domestic animals: "The quaggas even came among our cattle as they were grazing, and fed quietly with them; a proof how little shy they are in a place where they are scarcely ever pursued."[67] Harris similarly reported that a quagga foal, "neighed and frisked by the side of the horses for a considerable time, before it discovered its mistake."[68] This account accords with other reports that quagga foals that had become separated from their herds ran alongside horses.[69] This behavior would have made it relatively easy to capture quagga foals if people had wanted them. Selous had similar observations of other subspecies of plains zebras, "They are very easily caught when young, and soon become quite tame. When very young, if one gallops in between a foal and its mother, it will sometimes follow one's horse right back to camp."[70] Capturing a foal under these circumstances would seem to be relatively straightforward.[71] Equally easy was the capture of a mountain zebra stallion that seemed to have been attracted by horses and entered a kraal with them.[72]

Wild animals are more easily caught than tamed, and zebras in general have earned a reputation for resisting human domination. Jared Diamond put zebras in the category of animals with a "nasty disposition," and observed that they bit and became more dangerous as they aged.[73] There is no doubt that wild quaggas were combative when vexed or threatened: a man who had chased a herd of quaggas and, to save expending a bullet, had tried to drive an exhausted animal over a cliff, was attacked by the cornered quagga which grabbed his leg, pulled him from his horse, and ripped off his foot.[74] In a similar situation, facing a wounded stallion that bit wildly when approached, Lichtenstein prudently shot the animal.[75] A quagga's kick fractured a man's skull, and the bite of another tore the fingers off a man's hand.[76] Farmers made use of this ability of quaggas to defend themselves against predators by keeping them with their farm animals at night. Quaggas knocked hyenas to the ground, bit them and trampled them with their hooves.[77]

Diamond acknowledged that zebras had pulled Lord Rothschild's carriage and had been "tried out as draft animals" in South Africa, but these mentions gloss over the extensive examples of their use in draft.[78] Olaf Mentzel, a German who served the Dutch East India Company from 1732 to 1741, wrote that in the Cape Colony, "Some colonists have taken the trouble of taming some of these animals and harnessing them to wagons alongside their horses."[79] Gordon also recorded that farmers used quaggas to pull wagons, and observed, "They pull well, they are tough and strong, but vicious, biting and kicking, but it [sic] is considered more tame by nature than the [mountain] zebra."[80] Sparrman was an enthusiast about the possibility of taming quaggas for traction:

The quagga I saw here, having been caught when it was very young, was become so tame, that it came to us to be caressed. It was said never to be frightened by the hyaena, but, on the contrary, that it would pursue this fierce animal, whenever this latter made its appearance in those parts; so that it was a most certain guard for the horses, with which it was turned out to grass at night.

That these quaggas might be broken in for the saddle or harness, I have not the least doubt; as just before my departure for Europe, I saw one driven through the streets in a team with five horses . . . Quaggas or zebras, properly tamed and broke in, would, in many respects, be of greater service to the colonists than horses: as, in the first place, they are more easily procured here; and next, being used to the harsh dry pasture, which chiefly abounds in Africa, they seem to be intended by nature for this country . . .[81]

Barrow agreed about the potential of quaggas; he observed that they were, "well shaped, strong limbed, not in the least vicious," and were "soon rendered by domestication mild and tractable."[82]

Attempts to use mountain zebras in the Cape Colony were less successful. Sparrman recorded an unfortunate attempt of "a wealthy burgher" who, believing that he had tamed some mountain zebras,

Was absurd enough to take it into his head to harness them all to his chaise, though they were not in the least accustomed either to the harness or yoke. The consequence was, that they directly ran back into the stable, with the carriage and their master in it, with such prodigious fury, as to deprive him and every one else of all desire to make any further trials of this kind.[83]

Quaggas were already extinct when Tegetmeier and Sutherland made an enthusiastic case for domestication of plains zebras, but what they wrote shines light on what might have been done with quaggas. They reported that a hunter on horseback had lassoed eight "half-grown wild zebras" in the early 1890s.[84] After a month's training, people could harness four, although the remaining four required a longer period. People paired each of the trained plains zebra with a mule as part of a team of ten led by two horses that pulled a stagecoach (Figure 4.4) owned by James Zeedesberg

Figure 4.4. Led by two horses, four plains zebras paired with mules pulled this stagecoach in the 1890s. Plate opposite page 55 in Tegetmeier, Horses, Asses, Zebras.

in what is now Limpopo Province in South Africa.[85] An observer, noting that the plains zebras never kicked and that they pulled well, described their performance enthusiastically: "The zebras, when inspanned (harnessed to the coach), stand quite still and wait for the word to go, they pull up when required, and are perfectly amenable to the bridle, and are softer mouthed than the mule."[86]

A benefit to domesticating zebras in Limpopo Province and in the tropical regions to the north is that this warmer wetter environment provided suitable habitats for biting insects that transmit trypanosomiasis and African horse sickness, a viral disease. The latter is often fatal for horses, donkeys, and mules. It infects zebras, too, but typically persists in them for only a few weeks after infection and does not appear to be fatal.[87] As zebras can serve as reservoirs for the virus; an insect vector that drinks their infected blood can then transmit the disease when it bites horses, donkeys, and mules. Consequently, the advantages of using quaggas and other zebras for traction and livestock protection would have been offset by the drawback that they can harbor a fatal disease.

Although quaggas were harnessed, Barrow observed, "few have given themselves the trouble of turning them to any kind of use."[88] William Somerville, who saw many quaggas during the 1801–02 Truter–Somerville expedition into the interior, provided an explanation:

The Quacha generally walks or canters – but his quickest pace is short of the speed of a horse. The quacha is easily tamed if caught when young, and is a very strong animal, a team of them were broke in by a farmer of my acquaintance, he informed me that in the waggon 6 of them could draw more than 6 horses – but they are easily fatigued if driven at a quicker pace than a walk.[89]

Somerville's observations are consistent with other reports that horses could outrun quaggas, and points to a reason their use in traction did not become commonplace – they were prone to fatigue. Other harnessed plains zebras were similarly reported to be "lacking in stamina and endurance."[90] Plains zebras' flightiness was another problem: "It is their tendency to give way to panic that makes their chance of being domesticated somewhat unlikely – the flutter of a bird's wing, the breaking of a twig, may cause them to wheel around and gallop off."[91] It is not simply a matter of nastiness that keeps zebras out of the harness.

Rather than breaking zebras for harness, another possibility was to use them to breed hybrids that would combine the manageability of a domestic equine with a zebra's resistance to diseases such as trypanosomiasis and African horse sickness. People have mated zebras with either horses or donkeys to give hybrids, sometimes termed "zebroids," "zorses," or "zebdonks." These practices have a long history: in 1773 an ass stallion was painted with stripes to persuade a "shy" zebra mare to accept the ass as a mate.[92] Generally, however, this subterfuge was unnecessary: keeping a stallion and a mare in the same enclosure would achieve the same outcome. In 1804, Lichtenstein recorded that a Cape Colony farmer kept a tame quagga stallion with horses to improve the breeding stock.[93] Sometimes, a zebra would join a herd of horses and a zebroid foal would be born.[94] James Cossar Ewart, a professor at Edinburgh University who had raised several horse–zebra hybrids, was of the opinion that, "With time and care most of them could be trained to any kind of work," and compared these hybrids favorably with mules as being safer and more manageable.[95] In fact, zebdonks were used alongside mules to pull Zeedesberg stagecoaches in southern Africa.[96]

Valuing Quaggas

The biologist Cuthbert John Skead speculated that "Perhaps the quagga was its own enemy in that it served no really useful purpose" except as food for Khoekhoe and to make bags and leather strips from their hides.[97] This appraisal was precisely the "narrow utilitarianism" decried by the German zoologist Lutz Heck.[98] Heck argued that humans should feel happiness at the very existence of wild animals, and that the wonder of their being was reason enough for creatures to live – an opinion held by many present-day animal lovers. Stephen Kellert, who wrote extensively about biophilia, would have classified this viewpoint as "aesthetic" in his typology of "values of life."[99] Several other values in Kellert's classification could also have been applied to quaggas: ecologistic–scientific, naturalistic, symbolic, humanistic, moralistic, and utilitarian, but these values, too, failed to prevent extinction.

Quagga were wild animals in an age when wildness was not much valued. Apart from the few settlers who hoped to substitute them for horses or used them to protect livestock, virtually no one considered them to be worth anything more than a ready source of leather or food for the poorest and least powerful people. The one exception, the one voice of appreciation for quaggas, was Thomas Pringle (1789–1834), a poet and abolitionist. It wasn't just quaggas that Pringle appreciated. He wrote about encountering a herd of over fifty elephants and having advocated for leaving them in peace:

I confess, too, that when I looked around on those noble and stately animals, feeding in quiet security in the depth of this secluded valley – too peaceful to injure, too powerful to dread any other living creature – I felt that it would be almost a sort of sacrilege to attempt their destruction merely to furnish sport to the great destroyer man; and I was glad when, after a brief consultation it was unanimously agreed to leave them unmolested.[100]

Pringle did not make the case that quaggas were as stately as elephants, but adopting an attitude corresponding to "moralistic" in Kellert's "values of life" typology, he wanted to let them live.

The *quagga*, whose flesh is carrion, and even whose hide is almost useless, might be permitted, one would suppose, to range unmolested on his native mountains; but man, when he has no other motive, delights to *destroy* for the mere sake of pastime. Thus the poor quagga, in the absence of better game, is often pursued for sport alone.[101]

Unlike van Riebeeck, who had seen quaggas as possible substitutes for the horses he needed, Pringle, writing in 1835, saw little practical value to quaggas. Equines did not have to be domesticated to matter to Pringle. This opinion could have done much to protect them from extinction, if it had been more widely held.

5 · *Quaggas Abroad*

I congratulate you about the Quagga taint.

Francis Galton[1]

In the eighteenth and nineteenth centuries, a good number of quaggas were exported from the Cape Colony to Europe.[2] The ships that brought domesticated equines to this small corner of the empire might have been the same ones that loaded a quagga or zebra for the return voyage. The confinement and motion of the long sea journey must have been even more harrowing for wild animals than they had been for the horses on the inbound journey. Once in the imperial metropole, sojourning quaggas had a very different life than they had in the Cape scrub and grasslands. They also bore a completely different significance for people. In Europe, there was no need for them to provide meat and hides, fight off predators, or survive African horse sickness. They were exotics, appreciated for their far-away origins and their strange appearance. Quaggas in Europe were taken up in specifically European concerns: as an object of exotic wonder, quaggas became a symbol of Europe's relations with the rest of the world; and as the subject of a theory of biological inheritance, telegony, they testified to concerns about the authenticity of fatherhood.

Exotic Quaggas and Wonder

Exotics had a widespread appeal to Europeans.[3] In nineteenth century Britain, people were well acquainted with exotic animals from reading, illustrations, museum specimens, and living animals – some even encountered on the streets.[4] Royalty and nobles often had private collections of wild animals, whereas public menageries and zoos provided ordinary people with excitement, wonder, and glimpses of a world beyond their own.[5] Some menageries had fixed locations, such as the

Exeter Exchange on the Strand in London, and others toured so that even people in rural areas were often familiar with exotic animals.[6] The earliest European menageries in Paris and Vienna were open to the public in the late-eighteenth century and were followed by the London Zoo in 1828; soon there were zoos in many major European cities. Zebras were prominent among exotic animals; in 1830, Windsor Royal Park featured four plains zebras, four mountain zebras, and three quaggas.[7]

There are records of at least thirty quaggas kept in Austria, Belgium, Britain, France, Germany, and the Netherlands from the late 1740s through 1883; Appendix 2 summarizes this information. Quaggas were sometimes moved from one location to another, with the most notable example being the stallion belonging to Louis XVI, which lived in the menagerie at the Palace of Versailles and which fared better during the French Revolution than some other Palace residents, to end his days in the Paris botanical gardens.[8]

A quagga mare at the London Zoo from 1851 to 1872 was photographed by Frank Haes in 1863 and 1864, and by Frederick York in 1870.[9] All five photographs show the same bleak enclosure; but the bars of the cage are prominent in Haes's 1864 photograph, and the imprisoned state of the mare is further emphasized by the man staring in at her (Figure 5.1). From 1858 to 1864, this mare lived with the stallion presented to the London Zoo by Sir George Grey, Governor of the Cape Colony, but the possibility of foals born in captivity ended when the stallion injured himself so badly in dashing against the walls of the enclosure that he was put down.[10]

If quaggas in zoos generated interest, it is easy to imagine how excited people were to see them serving as draft animals in the streets where they were "as subservient to the curb and whip as any well-trained horses."[11] Two quaggas pulled the carriage of Sheriff J. W. Parkins, one of the officials whose responsibilities included maintaining the Royal Menagerie at the Tower of London.[12] Perhaps he viewed taking the royal quaggas out for a spin as one perk of being a sheriff. Jacques-Laurent Agasse painted one of these animals in 1817, and the skulls of both quaggas were preserved at the Hunterian Museum of the Royal College of Surgeons, London, until they were destroyed by bombing in 1941. Another instance of a quagga pulling a carriage led to this expert first-hand testimony from Charles Hamilton Smith, a naturalist and horseman, that quaggas were, "unquestionably best calculated for domestication, both as regards strength and docility."[13]

Figure 5.1. Mare at London Zoo photographed by Frank Haes in 1864. Courtesy of the Zoological Society of London

Important questions about the lives of quaggas abroad remain unanswered, especially regarding their relationships with other animals if they were housed together and with the people who interacted with them. The quagga mare kept at Kew Palace, London, showed a range of behavior: she "seemed to be of a savage and fierce nature: no one would venture to approach it, but a gardener in the Prince's service, who was used to feed it, and could mount on its back."[14] The quagga's aggressiveness probably resulted from ill-treatment, for Edwards records that someone had fed her tobacco wrapped in paper which she ate, but her tolerance of the gardener who cared for her is notable and provides support for the sentiment that some quaggas were manageable if handled correctly. Had there been similar mistreatment of the quagga mare at the Exeter Exchange in the 1820s, which was reported to be "not very docile"?[15] Apart from these few examples, the details of the quotidian lives of other quaggas are lost. They must have been cared for as other equines, and without reports to the contrary, they might have behaved similarly, when treated appropriately by expert caregivers.

One Quagga and Paternal Anxiety

One quagga stallion in nineteenth century Britain became well known in connection with how people understood heredity.[16] This stallion was owned by George Douglas (1761–1827), the sixteenth Earl of Morton, and was mated with a virgin horse mare that was seven-eighths Arabian, giving rise to a filly with "very decided indications of her mixed origin."[17] Both the mating and the outcome were unexceptional as horses and donkeys were mated with quaggas or other plains zebras to produce partially striped hybrids with curiosity value, or in attempts to produce equines that might be more resilient to tropical conditions.

Later, the seven-eighths Arabian mare, then in the possession of Sir Gore Ouseley, was mated with a pure-blooded Arabian stallion. Here the story takes an unexpected twist as the resulting one-year-old colt and the two-year-old filly were said to bear resemblances both to their Arabian horse parents and to the quagga with whom their mother had previously been mated. Lord Morton, a Fellow of the Royal Society, reported that both offspring had stripes and a similar color to the quagga, and he included a note by Dr. Wollaston attesting to the accuracy of his descriptions.[18] The foals bore upright stiff manes that differed from the lank manes of Arabian horses, and the foals' groom claimed that the manes were always upright. Lord Morton summarized his report to the Royal Society: "They have the character of the Arabian breed as decidedly as can be expected, where fifteen-sixteenths of the blood are Arabian; and they are fine specimens of that breed; but both in their colour, and in the hair of their manes, they have a striking resemblance to the quagga."[19] Morton concluded his account by emphasizing: "the extraordinary fact of so many striking features, which do not belong to the dam, being in two successive instances, communicated through her to the progeny, not only of another sire who also has them not, but of a sire belonging probably to another species; for such we have very strong reason for supposing the quagga to be."[20]

Morton's descriptions were well received, as evidenced by the Royal College of Surgeons commissioning the famous animal artist Jacques-Laurent Agasse to record the details. In 1821 Agasse produced a series of six paintings of the equines including "the first sire" (Figure 2.9), the female hybrid of the quagga and the Arabian mare, the Arabian stallion, described as "the second sire," the colt and filly born to the Arabian stallion and mare, and the Arabian mare with a young foal.[21] The paintings show that the colt and filly have stripes on their necks, shoulders and legs.[22] Leg stripes, of course, are absent in quaggas, which should have promoted some reflection about whether they were inherited from the quagga stallion, but this incongruity was not commented upon. Also unheeded was the absence in the colt and filly of the ventral stripe running along the middle of the abdomen, normally present in quaggas.[23]

The apparent influence of the quagga sire on subsequent offspring of the horse mare seemed to provide conclusive evidence for a belief termed "telegony" by the biologist August Weismann, who merged Greek words meaning "at a distance" and "offspring."[24] Writing in 1844, Charles Darwin applied Lord Morton's observations more generally:

When the dam of one species has borne offspring to the male of another species, her succeeding offspring are sometimes stained (as in Lord Morton's mare by the quagga, wonderful as the fact is) by this first cross; so agriculturists positively affirm is the case when a pig or sheep of one breed has produced offspring by the sire of another breed.[25]

In the absence of scientific evidence, the means by which the quagga could have influenced the subsequent offspring of the horse mare was open to speculation. To provide an explanation, Darwin invoked a mechanism involving "pangenes" and so used the term "pangenesis" to describe this form of inheritance. Darwin hypothesized that "pangenes" or "gemmules" were minute particles that could divide and transmit information from parents to offspring and even to subsequent offspring. Darwin believed that cells produce gemmules that circulate through the body tissues, "They are supposed to be transmitted from the parents to the offspring, and are generally developed in the generation which immediately succeeds, but are often transmitted in a dormant state during many generations and are then developed."[26] According to this thinking, the quagga would have transferred gemmules to the horse mare during mating and these would have been transmitted from the mare to the foals sired by the Arabian stallion where they would have given rise to striping and manes resembling those of the quagga stallion, the first sire. Darwin concluded: "Hence there can be no doubt that the quagga affected the character of the offspring subsequently begot by the black Arabian horse."[27]

The supposed influence of the first sire was an old belief that dated back at least to Aristotle, and the heart of the matter lay in uncertainties about human – not horse – fatherhood. *A Text-Book of Human Physiology*, published in 1888, noting that telegony was "incontestable," stated, "A woman may have, by a second husband, children who resemble a former husband."[28] This illusory correlation, of course, ignored the many counterexamples in which offspring did not resemble the first mate of their mother. The horse foals that seemed to possess characteristics of the earlier quagga mate of the mare dramatically provided a confirmation for a host of anxieties about offspring of women who had practiced serial monogamy or who had not held perfectly to their marriage vows.

Lord Morton's quagga and its supposed effect on future generations was well known in the nineteenth century and influenced the thinking of not only animal breeders but also prominent scientists and intellectuals.[29] Telegony influenced thinking in nineteenth-century southern Africa both in the context of sexual relations between settlers and Africans, and in animal breeding where mating a horse mare with a donkey to produce a mule was avoided because the mare was viewed as "polluted" and would produce inferior offspring if later mated with a horse.[30]

The playwright August Strindberg, with his own concerns about the paternity of one of his children, was well aware of the theory of telegony and in Act 2 of his 1887 play, *The Father*, he cites Lord Morton's observations to conclude that it is difficult to establish paternity based on the appearance of the offspring:

CAPTAIN.	Is it true that you obtain striped foals if you cross a zebra and a mare?
DOCTOR (astonished).	Perfectly true.
CAPTAIN.	Is it true that the foals continue to be striped if the breed is continued with a stallion?
DOCTOR.	Yes, that is true, too.
CAPTAIN.	That is to say, under certain conditions, a stallion can be sire to striped foals or the opposite?
DOCTOR.	Yes, so it seems.
CAPTAIN.	Therefore, an offspring's likeness to the father proves nothing?
DOCTOR.	Well ...
CAPTAIN.	That is to say, paternity cannot be proven.[31]

Strindberg was using the Captain's words to express anxiety about the powerlessness of the legitimate husband. A child conceived in an extramarital affair could bear resemblance to the husband who had been the wife's earlier lover and the source of the "gemmules" responsible for this similarity. Consequently, the adultery might not be apparent – a situation encapsulated in an old Latin saying, *Filium ex adultera excusare matrem a culpa*, which translates roughly to "The son exonerates his mother from adultery." Daughters were also subjects of concern, because of the belief that a woman's first lover could – even in the absence of pregnancy – deposit in her body gemmules that could contaminate a family's bloodline for ever. It was no accident that Darwin used the word "stained" to refer to the effects of Lord Morton's quagga on subsequent offspring, and that his cousin, the eugenicist Francis Galton, similarly wrote of the "Quagga taint."[32]

It was a no-win situation for husbands. Not only did they worry that the extramarital conceptions of unfaithful wives would go undetected, a wife's earlier sexual history could produce a child whose appearance was indebted to a long-previous lover. The latter situation was the theme of Émile Zola's novel, *Madeleine Férat* (1868) in which the protagonist bears a daughter who resembles her first lover, James – a close friend of her husband. Zola summarized the dilemma of her situation: "Although she had received the seeds of maternity from her husband, the young wife had given to her child the features of the man whose imprint she could not get rid of. There was no doubt that James's blood had had a large share in her impregnation; he who had made the virgin a woman was the first father."[33] Zola did not specifically mention Lord Morton's quagga in his novel, but – like other intellectuals of this time – he would have known about this case and used it in constructing the plot.

Lord Morton's observations influenced thinking about heredity for much of the nineteenth century but was disproved by a thorough series of breeding experiments and observations. With quaggas extinct, James Ewart used a stallion named Matopo which belonged to a different subspecies of plains zebra, *Equus quagga chapmani*. He mated Matopo with the mares of several horse breeds, which he subsequently mated with stallions of the same breed as the mare. Ewart examined the animals' foals, and concluded, "Neither in the colour, mane, tail, hoofs, call, make, nor yet in the disposition of the subsequent foals already bred, is there any suggestion of either the previous zebra sire or of their hybrid kindred."[34] Ewart exhibited Matopo and his offspring in 1900 at the annual show of the Royal Agricultural Society of England where they attracted much attention, although as Harriet Ritvo has noted, their significance regarding disproving telegony was downplayed by both the press and Ewart himself.[35]

Ewart's well-designed series of breeding experiments and careful observations of the animals born was important in discrediting telegony, but he went further by critically examining historical evidence. He noted that Agasse's painting of the filly showed that her mane lay to one side as is typical for horses, and was not standing upright as in zebras – showing that an essential piece of Lord Morton's testimony was flawed: the manes of the foals were not always like those of zebras as incorrectly claimed by the foals' groom.[36] Curiously, no one else appears to have commented on the discrepancy between Lord Morton's account and Agasse's painting.

Some people viewed telegony as a mental, rather than a biological process, positing that a female having seen some unforgettable image could somehow shape her offspring accordingly. One proponent, Sir Everard Home, viewed the birth of striped foals to the Arabian mare who had been mated with Lord Morton's quagga as "one of the strongest proofs of the effect of the mind of the mother upon her young that has ever been recorded." Ewart took a wry delight in debunking this idea. The quagga stallion, he quipped sarcastically, "had produced a profound and lasting impression on her nervous system; that she, as it were, had never quite succeeded in getting the quagga out of her mind."[37]

While systematically disproving telegony, Ewart also observed occasional striping in horse offspring of stallions and mares that had never been in contact with zebras, which led him to the conclusion that the ancestors of horses had stripes and so the occasional appearance of striping represented a reversion to an ancestral state – the phenomenon of atavism. Ironically, Darwin had also observed in *The Origin of Species* (1859) that striping sometimes occurred in equine hybrids and had speculated that the ancestors of extant equines were striped; nonetheless, he interpreted the striped offspring of the Arabian horses as resulting from telegony, not reversion to an ancestral condition.[38]

Ewart's experiments put to rest discussions of the effects of Lord Morton's quagga on subsequent generations, and in 1900 Gregor Mendel's experiments on inheritance in peas were rediscovered and, supported by experiments from the laboratories of Hugo de Vries, Carl Correns, and Erich Tschermak, the modern science of genetics was born. Ewart's straightforward breeding experiments could, of course, have been carried out much earlier, and nineteenth-century science would have been spared the fruitless distractions of telegony.[39] Lord Morton's observations, however, were given such credence because they appeared to afford visible scientific proof to support pre-conceived ideas.[40]

European Breeding

As the case of Lord Morton's quagga shows, Europeans frequently put quaggas in the company of their equine cousins: horses and donkeys. The exoticness of quaggas stood out when they were exhibited, portrayed, trained, driven, and mated with other equines: a mare with a donkey, another mare with an onager – the Asian wild ass – and a stallion with a horse mare.[41] All these matings produced live hybrid offspring, but probably the progeny were sterile.

Only relatively rarely did Europeans put quaggas together, to share the company of their species and possibly reproduce. There are three instances of a stallion and mare living together: in Windsor Great Park in the 1830s, at the London Zoo from 1858 to 1864, and at the Antwerp Zoo in Belgium in 1861. In the latter case, two foals were born; this is the only known instance of quaggas born in captivity.[42] There were no other births and the fate of the two Antwerp Zoo offspring is unrecorded. Captive breeding in Europe could have achieved survival even as the African population of quaggas declined, but unfortunately there was little interest in this course of action.

6 · *Extinction*

That an animal so beautiful, so capable of domestication and of use, and to be found not so long since in so great abundance, should have been allowed to be swept from the face of the earth, is surely a disgrace to our latter-day civilisation.

Henry Bryden[1]

Writing in 1875, the eminent biologist Philip Lutley Sclater (1829–1913) noted the declining populations of large African animals, but asserted, "The continent is too vast to admit of their extermination."[2] In one case, he was wrong: less than a decade later, quaggas ceased to exist. Sclater, Secretary of the Zoological Society of London, would have known about the extinction of island species such as dodos. The reason he seemed more sanguine about African animals probably resides in the phrase "too vast," which hints at extensive ranges that included places out of reach of hunters and settlers. His statement did not hold for quaggas, whose range was limited – akin, in fact, to being on an island, an island over-swept with a swamping tsunami powerful enough to drown out life.

The Wave of Extinction

No reliable estimates exist for the size of quagga populations before the establishment of the Cape Colony. A figure of "hundreds of thousands" in the eighteenth century seems reasonable, but even this approximate figure has no evidence to support it.[3] The South African biologist John Skinner wondered whether quaggas were, in fact, less abundant than generally supposed.[4] A small population and a relatively limited natural range further diminished by farms that excluded quaggas from grazing and water sources would certainly have hastened extinction.[5]

Quaggas disappeared along a wave of frontier settlement from the Cape. Eighteenth- and nineteenth-century observers documented their rarity and subsequent disappearance in an ever-widening area whose epicenter was Cape Town. Robert Gordon observed the beginning of this loss, noting in 1777 that quaggas "are no longer found near the Cape."[6] It was only when he had reached the interior in 1812 that William John Burchell observed, "a sight we had never before seen during our whole journey," reporting that, "As we advanced we saw at a distance around us, in every quarter, innumerable herds of *wild animals*, quietly grazing like tame cattle. Quakkas, springbucks, kannas [elands] and hartebeests on all sides."[7]

Accounts of other travelers in the late eighteenth and early nineteenth centuries describe the loss of wild animals: shooting by farmers was "rapidly diminishing" populations of elands and gemsboks in the Karoo; mountain zebras had become rare and blue antelopes were "almost extirpated."[8] Recognizing the seriousness of the situation, the British Governor of the Cape Colony, Lord Charles Somerset, issued a proclamation in 1822 "to guard against the total destruction of Game, in this Colony."[9] This proclamation promised rewards for killing "Vermin, or other noxious Animals" (the listed examples of such animals include predators of game), required licenses to kill other game (including zebras), imposed a closed season for hunting (July 1 to November 30), and levied penalties on transgressors. Hunting licenses required use of "Gun or Dog" whereas "Net, Snare, Spring, or other Engine" was banned – a provision that excluded some traditional African hunting techniques. However, the challenges of enforcement made this legislation ineffectual.[10]

The killing continued: quaggas had "almost totally disappeared" from the Albany District surrounding Grahamstown by 1821 and were gone from the Karoo by about 1860.[11] Figure 6.1 maps the last dates when quaggas were seen at particular locations or when they were reported to have been extirpated from those locations.

Thomas Baines encountered quaggas during his journey into the interior, but it was only when he reached the Orange Free State in 1850 that he saw them and other animals in abundance:

On surmounting a little rising ground, the flats I mentioned before presented a panorama of life and motion that will never be effaced from my memory. As far as we could see before us stretched a line of gnoos, quaggas, blesboks, harte-beestes, and behind, across a valley of six or seven miles in breadth, the same line

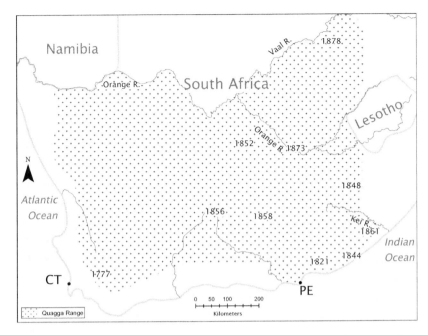

Figure 6.1. The progression of extinction. The stippled area shows the range of quaggas in the eighteenth century, and the dates indicate when quaggas were reported to be extinct or scarce at that location. This figure by the author was prepared by Bruce Boucek and Camille Tulloss; it is modified from a map published in Heywood (2015) and is reprinted courtesy of Kronos

was continued more or less dense, it is true, but still unbroken, till the rear of the immense herd was lost behind the rising ground in the distance.[12]

The quaggas Baines saw in such numbers had survived at the north-eastern part of their range and would be hunted out over the next thirty years or so. The last quaggas lived in the Orange Free State, south of the Vaal River.[13] They did not occur north of the Orange and Vaal rivers, and so only when people explored this area did their extinction become clear. The closely related Burchell's subspecies was also extirpated in the region where the last quaggas were shot, but it survived because its extensive range extended west into South-West Africa (present-day Namibia) and east into Zululand (now part of KwaZulu-Natal).

Even if the possibility of extinction had been appreciated and laws protecting quaggas had been legislated, enforcement would have been very difficult and it is unlikely that such efforts would have deterred rapacious hunters from killing the last quaggas. The survival of Cape mountain zebras and blesboks on southern African farms as described in the Introduction points to one possible solution. Breeding of animals in zoos overseas could also have been effective. Sadly, neither of these approaches was implemented.[14]

Skead, who had carefully studied the records of quaggas, believed that they were "still reasonably common" in the Orange Free State in 1873, whereas Bryden asserted they were extinct there by that year.[15] 1878 has been suggested as the year quaggas went extinct in the wild, though that is guesswork as it is difficult to assign an exact date to the final disappearance of an animal that had already become rare and that lived in a vast, under-populated region.[16] In any event, the quagga mare that died as an endling in the Amsterdam Zoo on August 12, 1883, probably outlived the last wild quaggas. How could Sclater have been so wrong? Why did one subspecies become extinct when other zebras survived? Quaggas were lost because their world was not the one envisioned by Sclater: it was not extensive or isolated.

Blaming Bystanders and Victims

The force that tipped the shrinking quagga population over the edge into extinction was hunting; however, some apologists have attempted to deny this conclusion by displacing the blame onto bystanders and even the victim. These hollow explanations include predation by lions, which casts humans in a minor role, as well as hunting by Africans, which excuses white colonists.[17] Sir John Kirk's conclusion about the causes for the depletion of wild animals is applicable here, "It is wonderful how little effect natives with spears, traps and arrows have on game in a country, and how suddenly it disappears before the gun and the rifle."[18]

Accusing the indigenous population of causing extinction not only obfuscates the truth but ignores the fact that they hunted for subsistence, which was itself imperiled by the large-scale hunting of colonists and by legislation which made illegal traditional African methods of hunting. Viewing lions as responsible for extinction is particularly disingenuous as settlers hunted both lions and quaggas: indeed, Cape lions (*Panthera leo melanochaita*) became extinct at much the same time as quaggas.[19] Both predators and indigenous people undoubtedly had an effect on quagga populations; however, this loss was sustainable over many millennia during which time droughts and loss of grazing land would have also taken their toll. It was large-scale hunting with firearms that brought about extinction.

Bryden compared the pace of horses and quaggas and speculated that "It is possible that the somewhat slower pace of the true quagga may have led to the animal's comparatively early extermination."[20] Horses could outrun quaggas, but the key factor was, of course, not the top speed at which quaggas could gallop, but the rifles that were used to shoot them – often by hunters on horseback.[21]

The writer Frank Beddard suggested that quaggas bore responsibility for their fate because of their nature as "a kind of terrestrial mermaid" with fore-bodies of a plains zebra and hind-bodies of a donkey; consequently, he blamed their extinction on being "intermediate between the two races."[22] This reasoning reflects a commonly held abhorrence of miscegenation and hybrids present in such diverse sources as the Old Testament (Leviticus 19:19) and Shakespeare, who described hybrids as "nature's bastards" in *The Winter's Tale*. The South African eugenicist Gerhardus Eloff also held quaggas liable for their doom, declaring that they had an impoverished gene pool and had failed to adapt to environmental change.[23] Left unexplained in both arguments was why the perceived deficiencies should have led to extinction at precisely the time that it did. Lurking behind these views was the erroneous thinking that extinction resulted from some shortcoming on the part of quaggas – as hinted at by the use of "schlemiel" in the title of Reinier Spreen's book, *Monument voor de Quagga: Schlemiel van de uitgestorven Dieren*.

Such spurious reasons for extinction are not unprecedented. Thylacines, *Thylacinus cynocephalus*, marsupial mammals that lived in Tasmania, Australia, and were known as Tasmanian Tigers, were also blamed for their own extinction – being viewed as oddly formed and unintelligent creatures.[24] The extinction of passenger pigeons (*Ectopistes migratorias*) in North America, also caused by hunting on a large scale, was met with the same denial of human agency: some people believed their disappearance resulted from mass drowning in the Great Lakes, the Pacific Ocean, or the Gulf of Mexico.[25]

Environmental Factors

Hunting caused the extinction of quaggas, but farming combined with environmental conditions may have hastened the process. One of these factors was a diminished availability of grazing, itself brought about by several changes. The proportion of grasses to shrubs in the Karoo has varied over the long-term, but the amount of grassland decreased beginning about the middle of the sixteenth century – a change that preceded farming by white settlers there.[26] The nineteenth-century migration of farmers and their grazing stock into the Karoo intensified environmental change and, by the late 1870s, overstocking, droughts, and locusts resulted in less palatable vegetation for sheep.[27] Changes in vegetation would have affected wild animals, too, and some suggest that the extinction of quaggas might have been accelerated by the effects of overgrazing and climate change on the vegetation of the Karoo.[28]

The decade of the 1870s brought further developments detrimental to quaggas: the increased use of fencing on farms that limited their access to prime land and water and that would have hindered their migrations.[29] Consequently, conditions had deteriorated even before the Karoo's "time of dread disease and drought" that killed thousands of farm animals beginning in 1877, and that dire episode, too, would have contributed to the extinction of quaggas if any survived by that time.[30]

Did Confusion about Identity Contribute to Extinction?

Reinhold Rau wondered whether a general inability to distinguish quaggas from other zebras and consequently ignorance about their numbers contributed to their "accidental disappearance."[31] He cited the survival of bontebok antelopes (*Damaliscus pygargus*), which were protected, and argued that quaggas could have been similarly protected if the gravity of their situation had been known.[32] This explanation does not hold, however, because other megavertebrates such as Cape warthogs (*Phacochoerus aethiopicus aethiopicus*), bluebuck antelopes (*Hippotragus leucophaeus*), and Cape lions (*Panthera leo melanochiata*) went extinct when there was no confusion about their identities.[33]

Nor does the case of Cape mountain zebras support Rau's argument: their shooting had been prohibited since the mid-eighteenth century, and the Game Law Amendment Act of 1886 should have afforded them some additional protection.[34] However, their numbers continued to decline into the twentieth century until less than 100 survived.[35] When Jan Kemp, Minister of Lands, was asked in 1936 to create a reserve for them, his reply, "No! They're just a lot of donkeys in football jerseys," reveals a casual official attitude to endangered species.[36] In 1937, a farm with six Cape mountain zebras was designated as a national park and so they were saved from extinction, but it was a very close call.[37] Absent this sort of initiative, however, there was nothing to preserve quaggas in the face of large-scale slaughter and public indifference. Rau himself, the visionary behind the rebreeding of quaggas, later forsook his opinion that their extinction was an "accidental disappearance." His ultimate conclusion was that, "The quagga became extinct through man's ignorance and greed. It wasn't a natural occurrence."[38] Rau also considered extinction "a dreadful mistake made by our forebears."[39]

Killing for Sport and Profit

The extinction of quaggas in the nineteenth century was hastened because of new connections, through British control, of the Cape Colony with the Global North. First was the establishment of commercial wool production in what is now South Africa's Eastern Cape Province. Second was the upsurge in big game hunting by leisured elites seeking trophies from the herds and flocks on the receding wildlife frontier. Third was the market demand, in Europe and the United States, for leather. Connections with global capitalist markets were intensified through the discovery of diamonds in quagga territory.[40]

The development of the Karoo and eastern Cape Colony more generally as a source of exports meant continued improvement in overland transport, ensuring that hides could be readily carried from the interior to ports and thence into the international market. Intensified pressure came from the migration of settlers into areas that had formerly been only sparsely inhabited. Among the early settlers, there were few full-time hunters. Rather, farming and hunting were intimately connected, so that most stock farmers were also hunters. Stock farmers who had found prime grazing locations when out hunting then settled on those lands, enjoying the benefits of grazing domestic animals and hunting wild animals, until the latter were either depleted or moved away.[41]

One such area was the Karoo, where expansion of farming not only brought wild animals closer to farmers' rifles but also resulted in a loss of access to water and grasslands that were now reserved for livestock.[42] With more people in the Karoo, more animals were shot for meat, hides, trophies, ivory, and sport, or because they were perceived as vermin that used grazing and water resources needed for domestic stock.[43] Hunting often proved to be more profitable than farming. Farmers, "shot game to support themselves, their families, and servants, and for the pure pleasure of hunting. But, so soon as they found a market for the skins of the game animals around them, they became mere hide-hunters, and shot, week in week out, for the mere value of the pelts."[44]

Hunting was such a part of life in the Cape Colony that in 1843 near Colesburg, where quaggas were still present, shots were fired almost every daylight hour.[45] Even a modest level of quagga hunting by farmers and others to obtain meat for their servants and hides for sundry purposes was probably not sustainable, but local farmers weren't the only ones shooting wild animals. The assault on wild animals in southern Africa intensified because of hunting expeditions by big-game hunters, mostly from Britain, which now claimed the Cape within its empire. Hunters such as William Charles Baldwin, Roualeyn Gordon-Cumming, William Cornwallis Harris, William Cotton Oswell, and Frederick Courteney Selous led treks that sometimes collected scientific information and supplied skins and skeletons to museums, but the scale of killing went far beyond this objective.[46] It might seem that these hunters were mainly interested in trophy animals and would have spared quaggas, but Harris reported shooting many of them.[47] Because it was the stereotypical food for servants, these hunting parties always had reason to shoot quaggas.

Killing was not strictly utilitarian. The excitement of the hunt also fueled the frenzy. Cornwallis Harris provides this insight, "From my boyhood upwards, I have been taxed by the facetious with *shooting madness*, and truly a most delightful mania I have ever found it."[48] In 1836 Harris, on medical leave from military service in India, landed at Algoa Bay of the Cape Colony (now in the Eastern Cape Province) and, accompanied by the hunter Richard Williamsburg, set out into the interior on a hunting expedition through southern Africa that netted a collection comprising a sable antelope skin and: "two perfect crania of every species of game quadruped to be found in Southern Africa, together with skins of the lion, quagga, zebra, ostrich, &c., tails of the camelopard [giraffe], and tusks of elephants and hippopotami, besides elaborate drawings of every animal that interests the sportsman, from the tall giraffe to the minutest antelope."[49] A man of contradictions, Harris waxed poetically, "There was something truly soul-stirring and romantic in wandering among these free-born denizens of the desert . . .," seemingly without contemplating that he was destroying what he treasured.[50]

Among the killing, one incident stands out: the hunt that Andrew Bain organized to celebrate the visit of Queen Victoria's son, Prince Albert. On August 24, 1860, local people were recruited to drive wild animals towards Bain's farm near Bloemfontein in the Orange Free State, where thousands were killed.[51] Thomas Baines's painting *The Greatest Hunt in History near Bloemfontein 1860* documenting this event shows stricken zebras, and the slaughter undoubtedly had a major impact on the remaining quagga population in the area.[52]

The final threat to quaggas seems most distant, but it was the largest, in terms of the number of people involved, the global scope of interests, and the amount of profit to be made: the global leather market. At the beginning of the nineteenth century, Britain was an important location for hide tanning and manufacturing leather goods using untanned hides (cow, oxen, and buffalo) from Europe and North and South America.[53] The hide trade in southern Africa grew in the 1850s, and by 1865 southern African hides were being exported to both Britain and the United States.[54]

At about the same time, the hide trade almost extirpated the American bison (*Bison bison*). These large herbivores had provided meat and hides for Native Americans but hunting by Euroamericans reduced an estimated population of 30 million to a few hundred individuals. Hunters harvested only the skins and tongues, leaving the meat for wolves, with later collection of bones providing income for some rural people. Undoubtedly, hunting by Native Americans took a toll – as did droughts, wolves, diseases, and loss of grazing land to domestic stock.[55] The bison's demise was, however, because of large-scale hunting for skins that were tanned and manufactured into leather goods. Many Native Americans had depended on bison for their sustenance, and their removal dealt a major blow to them and their way of life. This outcome was deliberate: Euroamericans wanted both the wealth of hides and the subjugation of native people. The latter had already been achieved in the Cape Colony and the Orange Free State, and so profit from hides was the key motive for killing quaggas.

In the 1850s, boot makers and consumers discovered that the skins of quaggas and zebras furnished superior leather that could be made into expensive boots.[56] Quagga skins were in demand in Britain, "English bootmakers made the discovery that the hides of the black wildebeest and the quagga, particularly the latter, made the finest quality of so-called porpoise hide, now much in request for walking boots."[57] This hide was also valued in the United States, as reported in the *New York Times*:

The markets were filled with skins which, when tanned, gave leather of a quality and excellence, never known before, but the origin of which, as the material was still sold under old names, purchasers never suspected. Hides of the zebra and quagga arrived in tens of thousands; and good as horsehide is for the uppers of first-class boots, these were even better.[58]

The 1871 discovery of diamonds in Kimberley also factors in extinction. By the following year, there were approximately 50,000 diamond miners in this area, which was not far from the last of the quaggas in the Orange Free State.[59] These miners had a major effect on wildlife and, "The quaggas suffered most of all, because their flesh was tasty, their hides strong and useful."[60] The wagons that carried supplies and people to the diamond mines of Kimberley might have transported quagga hides on their return journeys. The combination of flesh to fuel the extraction of precious stones and hides that were valued internationally made hunting quaggas vastly more attractive than when their meat had been given to slaves and laborers and their hides used to make low-value items. The buyers of diamond jewelry and boots made from quagga hides were doubtlessly unaware of the impact of their purchases on the environment and quaggas, but these effects were significant.[61]

The effects of the hide trade in southern Africa were substantial. Eventually, quaggas were extirpated from most of their range and survived only in the grasslands of the Orange Free State, but there was no respite. The evidence of carnage remained for decades; quaggas were already extinct when Bryden described this scene.

The bone-strewn plains of the Free State bear miserable testimony to the feverish haste and deadly methods of the skin-hunters of the past twenty years. The work was done with business-like skill and parsimony. Bullets were carefully cut out of the carcass for future use, and a sufficient number of skins having been prepared, the waggons were loaded up and slowly driven down country to the coast.[62]

One measure of the scale and results of the killing was that the number of wildebeest and zebra hides exported from southern Africa declined from 62,000 in 1873 to 7,000 in 1883.[63] By 1887, a report from London noted that instead of being delivered "in hundreds and thousands from South Africa, quagga hides now only appear in driblets of from ten to thirty at a time."[64] Undoubtedly, "quagga" here referred to zebras in general, but the hides of true quaggas would have been among the large numbers of zebra hides originally exported. Dead quaggas were valuable and, as noted by the *New York Times* in 1887, the live ones became extinct: "The quagga, the beautiful wild striped ass of South Africa, has suddenly ceased to exist. The bootmakers of London and New York wanted his skin for a particular kind of sportsman's boot, and he consequently passed away out of zoology ... Animals which when dead are exceedingly valuable contract a habit of dying."[65] There was no recourse against this onslaught: the 1822 proclamation by the Governor of the Cape Colony to protect game was ineffectual, and attempts, between 1866 and 1870, to pass legislation that would have protected wild animals, including quaggas, were unsuccessful.[66]

The 1880s represented a sea change for "game" animals in southern Africa. At the end of the decade, the game hunter Frederick Selous concluded his hunting expeditions.[67] In August 1883, the very month that the last quagga died in captivity, the Natal Game Protection Association was organized as a conservation group.[68] And on July 6, 1886, the Government of the Cape of Good Hope passed The Game Law Amendment Act, 1886. Section 4 of this Act reads: "No person, however, shall be at liberty to pursue, shoot, kill, destroy, or capture any elephant, hippopotamus, buffalo, eland, koodoo, hartebeest, bontebok, blesbok, gemsbok, rietbok, zebra, quagga, Burchell zebra or any gnu or wildebeest of either variety, without having obtained a special permission to that effect from the Governor . . ."[69]

The aim of the law was to preserve game animals for licensed killing by colonists who shot animals "cleanly" for sport. Like the 1822 proc-lamation by Lord Somerset, these regulations made it impossible for Africans who used traditional methods, such as pits and poison, to hunt legally for subsistence.[70] The irony that the Game Law Amendment Act was passed almost three years after the last quagga died in the Amsterdam Zoo has been the subject of frequent comment.

The extirpation of quaggas and plains zebras from the Cape Colony and Orange Free State came with a benefit for farmers: the danger of African horse sickness to domestic equines was lessened.[71] Likewise, the excessive hunting that removed almost all the zebras also reduced the populations of predators such as lions, hyenas, and wild dogs, and so the need for quaggas to protect livestock was reduced. No types of zebras were domesticated and quaggas became extinct. Such were the gains and losses in the changing world of late nineteenth-century rural southern Africa.

Regrets and False Hopes

"After-comers cannot guess the beauty been," the nineteenth-century poet Gerard Manley Hopkins wrote in grief over the destruction of a grove of trees.[72] These words apply equally well to the extinction of quaggas and the emotions their loss has stirred in South Africa and abroad. Expressions of regret focus on the shame of their disappearance and the hope that humans will learn lessons from their extinction.[73] Bryden voiced both these themes: the loss of quaggas was a "disgrace to our latter-day civilisation" that might "serve as some sort of warning to wanton and ruthless destroyers of game."[74]

Not everyone felt as strongly as Bryden. His contemporary, the hunter Frederick Selous, viewed their extinction as "deplorable" but, noting that they were the "most southerly form" of the plentiful plains zebra, he contrasted them with Cape buffaloes and rhinoceroses, which were these "highly specialized and most interesting creatures." Should Cape buffaloes and rhinoceroses be lost they would leave no species "which resembles them in the remotest degree."[75] In making comparisons between animals in terms of what species was the most important to exist, Selous was prefiguring the hard choices that face conservationists who must decide which organisms to protect when there are insufficient resources to save all.

Extinction evokes regret, but sometimes holds out the antidote of hope. In 2005, American birders were thrilled by reports that ivory-billed woodpeckers (*Campephilus principalis*), last reported in 1935, were present in an Arkansas swamp. The scientific community organized itself to weigh the evidence and birders swarming the region set off a minor tourism boom.[76] The confined space, good accessibility, and rapid news cycles made the search for the ivory-billed woodpecker a dramatic example of a broader phenomenon: dedicated searches for "Lazarus" species, named for Jesus Christ's friend whom he raised from the dead. Lazarus species do exist, most famously the coelacanth fish (*Latimeria chalumnae*) and the dawn redwood tree (*Metasequoia glyptostroboides*). These organisms, previously known only from the fossil record, have inspired searches for many others, sometimes in unlikely locations far away from the original habitat. Reports have put passenger pigeons, extinct in their native habitat in North America, in places such as Chile.[77] Thylacines, that became extinct in the 1930s, are still being sought – including at locations thousands of miles away from their last known habitats in Tasmania.[78] Quaggas were extinct in their well-known haunts, but this did not stop enthusiasts from searching for them in places they considered untamed enough to harbor them.

The searches began nearly a half-century after extinction. Writing in 1920, Frederick William Fitzsimons, Director of the Port Elizabeth Museum in South Africa, responded to a report that animals described specifically as quaggas – and not as mountain zebras or plains zebras –

lived in a hilly area of South-West Africa. "Knowing what a sensation the discovery of a few survivors of the once numerous race of Quaggas would cause," he asked hunters in South-West Africa to send him, "a skin and a skull of one of these alleged Quaggas with a view to settling the question one way or the other."[79] Old habits die hard.

James Drury of the South African Museum also received a report of quaggas and acted similarly: determined to investigate rumors in the 1920s of quaggas in the Cape Province, Drury was issued a permit to kill one animal – a decision which raises questions about both the effectiveness of the Game Amendment Act and his position on extinction. None were seen, but Drury's allies shot two Cape mountain zebras, a species that then numbered less than a hundred animals, in the Southerland District of the Karoo.[80] Fitzsimmons's and Drury's motivations are puzzling: finding quaggas would have brought them some fame, but it is uncertain whether they would have sought animals to stock reserves in South Africa – particularly because this approach still had not been implemented for Cape mountain zebras, which were facing extinction themselves. In any event, their tactic was more that of big game hunters than biologists.

Attention then returned to South-West Africa, where no one had reported quaggas. The reasoning seems to have been that mountain zebras occur as discontinuous populations scattered in South Africa and South-West Africa, and so it was possible that quaggas might also have a similar distribution and exist in under-populated areas where they could have escaped notice. However, a 1927 expedition to South-West Africa led by Guy Shortridge, Director of the Kaffrarian Museum (now the Amathole Museum) in King William's Town, Eastern Cape Province found none.[81] Nonetheless, some people continued to hope.

Under the headline, "Quagga Not Extinct. Herds in South-West Africa. Convincing Story from Namib" on June 18, 1932, a correspondent for the *Johannesburg Star* reported seeing a herd of eleven quaggas in South-West Africa, followed by a sighting of a further fourteen several weeks after that.[82] Two days later, however, the zoologist Austin Roberts pointed out, also in the *Star*, the obvious: how unlikely it was that quaggas, animals of the central plains of South Africa, would occur in the mountains of South-West Africa. Roberts suggested that the zebras in question were probably Hartmann's mountain zebra, a subspecies of *Equus zebra*. It was hard, Roberts explained, to distinguish between different sorts of stripes in poor light conditions.[83] Further reports of quaggas in South-West Africa prompted an official investigation in

1940 by C. L. H. "Cocky" Hahn, the Native Commissioner for South-West Africa, which came to a similar conclusion as Roberts's about Hartmann's mountain zebras and the difficulty of seeing stripes clearly.[84] The most reasonable explanation, however, was that the animals observed were plains zebras, since their bodies were reported to have reduced striping and a brown color, features that occur in some plains zebras in Namibia.[85]

Unconfirmed accounts of quagga sightings in South-West Africa continued into the 1940s. In the early 1950s, no less than three expeditions searched unsuccessfully in South-West Africa. First, in 1950, reports of quaggas in the Kaokoveld (a desert area on the northern coast of South-West Africa, present-day Namibia) inspired the explorer Quentin Keynes (1921–2003) to investigate. Keynes, a great-grandson of Charles Darwin and a nephew of the economist John Maynard Keynes, seems to have been drawn to quaggas as part of his fascination with extinct and rare animals, including coelacanths, dodos, giant sables, thylacines, and a spotted zebra.[86] Keynes and his companion did not find quaggas.

Next, in 1951, Bernard Carp (1901–1966), a Dutch industrialist who was the director of a distillery in Cape Town, led an expedition to the Fish River Valley. Carp had broad interests in biology and had led expeditions to several African countries searching for animal and plant specimens.[87] As a successful businessman, financial gain was probably not important, but finding quaggas would have brought esteem. He recognized that a search outside the original range of quaggas might be futile, but he gave some credence to the reported sightings of quaggas by Mr. Zelle, Conservator of the Windhoek Museum, and thought isolated herds might exist away from most human observers.[88] This view was not entirely unrealistic as the Fish River Valley was close to the Orange River, the northern limit of quaggas' range (Figure 0.2). Carp went as far as to offer a reward in South-West Africa for the skull and skin of a quagga. This gesture led to a succinct and scathing comment by C. L. Boyle, Secretary of the Fauna Preservation Society: "Should a few quaggas really exist what surer means can there be of making the animals' extinction certain?"[89] Publicity about Carp's offer prompted the authorities to insist that it be withdrawn, and he complied.[90]

Neither the failure of earlier expeditions to find quaggas, nor Roberts's observations that quaggas had been confined to the plains of southern Africa, nor Hahn's report that quaggas were absent from South-West Africa, prevented the National Parks Board from mounting in 1952, "a scientific expedition, to establish whether the quagga, long believed extinct, can be found." Under the headline, "To Search for Quagga in Remote Hills. Expedition to Leave Soon for S. W. Africa," a newspaper reported that: "The value of a quagga, dead or alive, would be enormous. American zoos offer "globular sums" for a live specimen. Even dead, the quagga has great value. One overseas museum, it is said, values its quagga skull at £25,000."[91] The expedition searched in a mountainous area of the Fish River Valley with the goal of determining whether quaggas still existed, and, if so, to attempt their capture. Accompanying a member of the National Parks Board were a Senator, a photographer and, ominously, a taxidermist, whose presence suggests that the statement, "Even dead, the quagga has great value," might have been taken literally.

What were the underlying motives for these expeditions in the mid-twentieth century, almost seventy years after the last quagga died? Although the newspaper report of the 1952 expedition stressed monetary value, the National Parks Board was involved in conservation in South Africa and its interests were undoubtedly scientific.[92] If quaggas had been found they would have been carefully observed and there may have been the intention to bring captured quaggas back to South Africa. This flurry of mid-twentieth-century searches put an end to any lingering hopes that quaggas were still alive, but at precisely this time an intriguing possibility emerged. A writer in the journal *African Wildlife* proposed that selective breeding of plains zebras might produce zebras resembling quaggas.[93] This idea was more realistic than finding survivors who had escaped hunters.

Was Extinction Inevitable?

Saving quaggas living in the wild was an almost impossible task given the value of their hides, the overall level of hunting, and the difficulties of enforcing laws that might have protected them. However, the survival of Cape mountain zebras and bonteboks on farms shows that private measures might have been enough to save animals sought by hunters. There might have been a similar outcome for quaggas if the foal nurtured in the 1850s by the farming family of Kamferskraal, Nelspoort in the Beaufort West District of the Cape Colony had survived and had been protected with a few of her kind.[94] Captive breeding would have been possible overseas where quaggas were valued as exotics, but unfortunately they were often maintained as solitary animals and, if bred at all, it was often with other species. The successful breeding at the Antwerp Zoo could have been repeated at other zoos and on private estates, but this opportunity, too, was lost.[95]

This chapter reveals as much about humans and their capacity for delusion and denial as it does about quaggas and their extinction. When quaggas were no more, humans saw them in other zebras, fantasized about their existence in remote places, and constructed reasons for extinction that avoided the obvious one. The following conclusion, however, is more credible. Quaggas inhabited a limited area, subject to environmental change in the form of droughts and loss of grasslands. By the time Europeans arrived, they may not have existed in vast numbers and there was constant loss to indigenous people and predators. However, although they survived environmental change and predation for millennia, they could not endure in the face of large-scale killing by nineteenth-century hunters, farmers, and others.

7 · *Afterlife*

Even dead, the quagga has great value.

Anon, 1952[1]

The epigraph refers to the worth accorded to taxidermy specimens and bones. These preserved tissues, however, gained additional value when their DNA initiated the discipline of paleogenomics, revised the classification of plains zebras, and provided the rationale for the Quagga Project. Biology was not the only sphere where quaggas experienced a reappearance: accounts and illustrations of them before their extinction have been joined in their afterlife by stories, poems, paintings, and a film. Moreover, quaggas epitomize two weighty current concerns: anthropogenic extinction and racial identity.

|Xam Narratives

This chapter begins with accounts of quaggas from Bushmen, people who knew the animals best. We know some |Xam narratives about quaggas from the work of the German philologist, Dr. Wilhelm Bleek, and his sister-in-law, Lucy Lloyd. Bleek arrived in the Cape Colony in 1855 with a plan to uncover the deepest relationships between the language families in Africa. He spent the rest of his life there, weaving his own family life into the research. His collaborator and successor in this research, Lloyd, lived with Bleek and his wife (and Lucy's sister) Jemima, in Mowbray (a suburb of Cape Town). In 1869, Bleek learned of 29 |Xam Bushmen held at the Breakwater Prison in Cape Town. The prisoners were from what is now the Northern Cape Province, near the towns of Brandvlei and Kenhardt; this dry, rocky, and inhospitable region is called "Bushmanland." In the mid-1800s, after two centuries of expansion of the Cape Colony and genocidal clearing of hunter-gatherers from the frontier, the region was the last refuge of |Xam Bushmen.[2]

By the mid-nineteenth century, Bushmanland was a backwater of the British-ruled Cape Colony, where poor Coloured and white settlers tried to emulate the sheep farming of better-favored regions. Stock farmers and hunter-gatherers make notoriously bad neighbors, and the settlers dealt with Bushmen by raising commandos against them. They cleared the territory through what a British official described as "a general practice of hunting Bushmen."[3] Many, many were killed. Some were merely arrested and transported to Cape Town, where the city-bound linguist Bleek recognized the research potential of learning their language and collecting their folklore.

In 1870, Bleek took several of the prisoners into his household, where he and Lloyd learned the |Xam language and recorded their narratives in 12,000 notebook pages of |Xam text with English translation. Despite the inequalities and dispossession that structured the research, Bleek, who died in 1875, and especially Lloyd, who continued the enterprise until 1884, became sensitive interlocutors of |Xam culture.[4] Through the Bleek–Lloyd archive of Bushmen folklore, the stories of a nearly extinct animal, told in a language that is also now extinct, are accessible online in a faded cursive text. The narratives throw light on Bushman ways: hunting techniques are described, domestic violence is condemned, and the fear of animals towards humans is explained.

Some narratives allude to an early time before |Xam existed and when present-day animals were humans. Dia!kwain's father told him that when quaggas were people, they looked like |Xam, and, as people, they associated with things that explain their color as animals. And so the stripes on a quagga's back resemble the reed mat she carried when she was a person.[5] In Dia!kwain's story, a quagga roasted seeds on the ground, gathered them up with both hands, placed them on her mat, sieved away the earth through the mat, and put the seeds in a bag. She carried the bag home and put the seeds on a flat stone with a concave upper surface. Holding a round stone in her right hand, she ground the seeds on the stone and then sieved them through her mat to remove husks from the flour.[6] As for the animal's dark coloration, it was said to derive from the food that the mantis gave to the quagga.[7]

Before people existed, baboons and quaggas were people, and both species feel that they are people – not game; because they were once people, the organs of these creatures still smell like those of people.[8] Despite their shared origins, in one story baboons killed a quagga girl who lived among them and served up her meat, claiming that it was a young gemsbok.[9] Baboons were not alone in being violent to quaggas in the household: a quagga married to a male jackal gave a piece of her own liver to their quagga child who had complained of hunger. But a tortoise stole the liver and gave some to jackals who reported to the male jackal that he "had married meat." The male jackal put bones covered with poison in their house, and these pierced the quagga's body when she lay down on them. As the poison took effect, she left the house and died beneath a tree where the jackals cut her up. The young quagga who was hiding in the tree saw what had happened. When the jackals had departed carrying the quagga's meat, the young quagga went to her equine grandparents who, hearing what the male jackal had done, trampled him to death.[10]

A similar account of a dysfunctional family features a quagga whose husband was a dog from a family of hyenas, jackals, and blue cranes. All praised the quagga's liver given to them by the tortoise and recognized its source. The Blue Crane tasted the liver and called: "Excellent! Excellent! my grandson !kuinssi/kauöken must truly have married a quagga. Wait! Wait! Let him come, that I may tell him about it; for it must be a quagga whom he has 'married!' And !kuinssi/kauöken came; they came to tell him that a quagga must be the one whom he had married; he should kill the quagga."[11] The dog did indeed kill the quagga and later he married her younger sister, but her parents forestalled further violence by trampling him to death.

Quagga meat was so valued that it was counterfeited, as when a lizard member of an earlier race entered a whirlwind where he cut off his own flesh, gathered it up, and carried it home to his wife who sliced it and served it – believing his claim that it was quagga's meat. One day, at his mother's request, their son accompanied the father on the hunt but, when a whirlwind drew near, the father instructed him to cover his head with a kaross [hide cloak] because quaggas were approaching. The son disobeyed and watched his father run into the whirlwind and emerge with what he claimed to be quagga's meat. The son's suspicions were confirmed when, surreptitiously, he sought the animal's spoor on the ground and found none. The son reported his findings to his mother, and she realized that they had been eating the father's flesh. Mother and son determined to return to their own people who hunted authentic quaggas.[12]

Marriage between herbivores and carnivores is bound to create problems. To prevent this situation and similar ones such as the lizard feeding his family with his own flesh, the Anteater, acting like the author of the Deuteronomic Code, laid down laws governing the appropriate food for animals and specifying that they must marry their own kind. In "The Anteater's Laws", quaggas were instructed to eat grass.[13]

Once quaggas had become animals, there had been a time when they and other game had been tame and, like domestic animals, had allowed people to stroke them. !xugen-ddi explained to his grandson Dia!kwain how a man called "Chaser-of-food" had brought this state of affairs to an end. He approached a quagga that slowed so he could stroke it but, rather than touching, he beat and kicked the animal. Since then, game animals have feared humans and fled from them.[14]

Hunting features prominently in |Xam narratives about animals. For example, the development of large flower buds on a particular species of *Mesembryanthemum* marked the appropriate time to hunt quaggas.[15] |Xam believed that their campfires attracted some moths that lived with animals. If these moths were burnt by the fire, it would augur which game would be caught and roasted on the fire. The appearance of a particular moth at the fire before the family's father returned from the hunt foretold that he would bring quagga meat.[16]

There was a cautionary tale that quaggas could be cunning: a sickened animal that had been shot with a poisoned arrow led a |Xam hunter to bushes where a lion was hiding.[17] In another story about hunting, the black wildebeest thwarted the attempt of the long-nosed mouse to hunt quaggas by blunting his arrowheads so that they would not penetrate the flesh and substituting his own entrails for the bowstring so that it would break when pulled. The wildebeest mingled with the quaggas and ran in front of them when the mouse scared them into flight. From this vanguard position, the wildebeest saw the mouse's shelter of bushes and trampled the mouse to death – so saving the quaggas.[18] That the protector of quaggas is a wildebeest is unsurprising, as |Xam people would have seen these animals grazing together. Furthermore, the narrative mirrors Bushmen hunting techniques: like the mice in the story, they caused animals to stampede, directed the flight of their prey by inserting into the ground ostrich feathers attached to sticks, and hid in bushes to ambush the animals.[19]

The animal subject of these stories and the language they were told in disappeared about the same time. The |Xam story tellers are also sometimes described in terms of extinction, but this does not do justice to human adaptability. Colonial rule crushed the language and lifeways of the |Xam, but, unlike quaggas, the people endured. When the |Xam were defeated and assimilated, their descendants survived as "Coloured" laborers on white-owned farms occupying the same land their ancestors roamed.[20] They speak Afrikaans and still tell the tales of their ancestors. In 2011, the researcher Jose Manuel de Prada-Samper heard some of the same stories told to Bleek and Lloyd and since then has documented a rich body of folklore.[21] People today don't tell stories of zebras, but de Prada-Samper thinks their knowledge of quaggas may live on in their connection to donkeys. In 2014, Katriene Swartz told him this interpretation about part of the Milky Way: "When you [are] in the dark night, my father and his people also pointed to us, when the Street [of Heaven] indeed turns, the Milky Way, there lies a donkey, a signal like a donkey. Either he lies down with his head that he lies towards the tip, or he had turned and then lies up."[22] Like many other South Africans, Swartz associated the donkey with the one Jesus rode on Palm Sunday,[23] but this is only one way that equines are imprinted on the landscape among the descendants of the |Xam. Paintings of horse-drawn wagons made in the late nineteenth century document the increased traffic to the diamond fields. At the same time, travelers recorded the first stories about ghost wagons traveling in the night. The increased hunting from diamond fields probably contributed to the extinction of quaggas. The descendants of the |Xam saw the carriage horses, cart mules, and dead quaggas and also noted the bones of trek animals and wild game that covered the open veld and roadsides. As the farm laborers of the Karoo tell stories of ghost wagons, they may be testifying to equine death, including that of quaggas.[24]

The Name Lives on

In the English-speaking world, the word quagga appears occasionally in crossword puzzles, alphabet blocks, and alphabet books. The names "quagga" and "kwagga," or much less frequently, "iqwarha," have been appropriated for a radio station, business enterprises, locations (including roads and a shopping center), an inspired pen name (Basil Roan Quagga), and fictitious characters (Quagga was a Wookie in *Star Wars*, and Kwagga Robbertse featured in the 1990 movie, *Kwagga Strikes Back*). A fictitious quagga chess piece figured in the 2003 Harvard-MIT Mathematics Tournament.[25] There is even *Quagga* software, which is a project fork of *GNU Zebra* software – bringing together in the virtual world animals of the veld. The fate of quaggas has been visited upon their namesakes: the journal *Quagga* was renamed *Endangered Wildlife*.

Extinction – or even a name change – is unlikely for the quagga mussel, *Dreissena bugensis*.[26] This animal with partially striped shells originated in Ukraine but was released into North American freshwaters, where it has become an invasive species whose prolific multiplication and filter feeding results in vast populations that both deplete the plankton needed to nourish fish, and clog screens and pipes in waterways. This mussel was so named to create an association with, but also to distinguish it from, the related invasive zebra mussel, *Dreissena polymorpha*, which also alters the ecology of its habitats.[27] Appropriately, quagga mussel shells are usually less heavily striped than zebra mussels. Finally, the dark brown stripes on a small shark resulted in its species name, *Halaelurus quagga*.[28]

Faux Quaggas

With quaggas extinct, some people attempted to profit from their relics – or even from claims of their continued existence. Guy Shortridge, Director of the Kaffrarian Museum, who had earlier unsuccessfully sought quaggas in South-West Africa, seems to have abandoned attempts to find the genuine article. He contacted the London firm of taxidermists, Messrs Edward Gerrard & Sons, who had mounted a hippopotamus named Huberta that had proved to be a major attraction of the museum.[29] Following this success, he now explored whether they could fake a quagga specimen. The taxidermists were equal to the challenge and ingeniously proposed the following creative taxidermy:

The markings we could reproduce fairly easily with stains and careful scorching. We believe the material to use for this would be the skin of one of your big donkeys ... A hot weather coat from a White Donkey (South African) or White Mule, if such a thing exists, would be most suitable, and one might say it was still a South African product.[30]

Shortridge never implemented this sleight, but whether he intended to pass off the specimen as a real quagga – or merely as a look-alike – is unknown.

Others continued to invoke the idea that quaggas still existed in remote locations. A man contacted Lutz Heck, formerly director of the Berlin Zoo, when he was visiting South-West Africa in the 1950s and offered to tell him where he could see a herd of quaggas for £70 and, as a gesture of good faith, promised to refund all expenses if the animals were not genuine.[31] Heck, who was himself not above crooked dealings – during World War II he had looted animals from the Warsaw Zoo to stock German zoos – declined the offer.[32] The attempted sale in 1974 of a mounted head said to have belonged to the last known quagga from the Eastern Cape was no more successful. The would-be seller had inherited this trophy from his father, who had shot the animal. The son seemed to have genuinely believed its provenance but measurements and photographs sent to the Natural History Museum in London showed that the head was that of a different subspecies of zebra.[33] It is curious that value was attached to the striped head of a zebra whose distinguishing characteristics are its unstriped legs, belly, and rump.

Quaggas in Culture

Whereas Bushmen featured quaggas in their stories, European writers found little creative inspiration in these animals: their presence in Thomas Pringle's poems, "Afar in the Desert" and "The Desolate Valley" are the exceptions.[34] European painters made beautiful depictions of quaggas, but sometimes as variations on well-established equine themes. Once they were lost, in an age of so much extinction, people had reason to reflect on quaggas and, reflecting, they produced art. In the twentieth century, portrayals of these animals often lingered on their fate.

"The Extermination of the Quagga" commissioned by the Austrian zoologist Otto Antonius and painted by the Austrian wildlife artist Franz Roubal in 1931 depicts a herd under fire from hunters on horseback.[35] Several quaggas have been shot: they rear back in pain or writhe on the ground. The perspective is horrifying, from within the herd, but close to the ground. Perhaps this slaughter is being seen by an animal that, too, has already been shot? A contemporary American artist, Walton Ford, also painted a dying quagga in 1998, this one knifed. The male, a beautiful animal, is collapsing gruesomely with both tongue and penis extended as his head hits the earth. At the edge of the quagga's body, the artist penciled in dates, its stripes serving as a timeline for the extinction of the animal.[36] Not all recent illustrations feature death: the sesquicentennial of the *Natura Artis Magistra* Zoo in Amsterdam was commemorated in 1988 with a stamp bearing the image of the quagga that lived and died in the Zoo as the last of her kind.[37]

The earliest known sculpture of a quagga is from the eighteenth century and was based on a living animal. Later three-dimensional depictions include work by Francis W. Eustis, who was a noted sculptor of horses, and Reinhold Rau, who painted models to raise funds for the Quagga Project.[38]

In the mid-twentieth-century, quaggas began to appear in fiction with the publication of "The Last of the Quaggas," which draws on the myth that some of these animals survived into the twentieth century and features a stallion born in 1904. Brought up on a farm, the quagga stallion roams afar in an unsuccessful search for companions and, with the aid of sympathetic farmers, eludes hunters eager to catch him.[39] The South African poet Roy Martin Macnab (born 1923) includes quaggas in depicting a world that is no more: in "Winged Quagga", they assume a mythical Pegasus status as he looks back to his childhood.[40] Quaggas share their lives with bacteria, protozoa, fungi, fire-flies, and glow-worms in The Quagga's World as Macnab recognizes that these small organisms were just as much part of quaggas' environments as were ostriches, gnus, and antelopes.[41]

"Elegy for the Quagga," by the contemporary poet Sarah Lindsay, imaginatively contrasted the 1883 eruption of Krakatau in a loud explosion followed by dramatic red sunsets with the quiet expiration of the last quagga that occurred a few days before:

Krakatau split with a blinding noise
and raised from gutted, steaming rock
a pulverized black sky, over water walls
that swiftly fell on Java and Sumatra.
Fifteen days before, in its cage in Amsterdam,
the last known member of *Equus quagga*,
the southernmost subspecies of zebra, died.
Most of the wild ones, not wild enough,
grazing near the Cape of Good Hope,
had been shot and skinned and roasted by white hunters.

When a spider walked on cooling Krakatau's skin,
no quagga walked anywhere. While seeds
pitched by long winds onto newborn fields
burst open and rooted, perhaps some thistle
flourished on the quagga's discarded innards.
The fractured island greened and hummed again;
handsome zebras tossed their heads
in zoos, on hired safari plains.
Who needs to hear a quagga's voice?
Or see the warm hide twitch away a fly,

see the neck turn, curving its cream and chestnut stripes
that run down to plain dark haunches and plain white legs?
A kind of horse. Less picturesque than a dodo. Still,

we mourn what we mourn.
Even if, when it sank to its irreplaceable knees,
when its unique throat closed behind a sigh,
no dust rose to redden a whole year's sunsets,
no one unwittingly busy
two thousand miles away jumped at the sound,
no ashes rained on ships in the merciless sea.[42]

Death also occurs in the poem "The Quagga," by Dennis J. Enright, which features the stallion that lived with its mate in the London Zoo from 1858 to 1864.

By mid-century there were two quaggas left,
And one of the two was male.
The cares of office weighed heavily on him.
When you are the only male of a species,
It is not easy to lead a normal sort of life.

The goats nibbled and belched in casual content;
They charged and skidded up and down their concrete mountain.
One might cut his throat on broken glass,
Another stray too near the tigers.
But they were zealous husbands; and the enclosure was always full,
Its rank air throbbing with ingenuous voices.

The quagga, however, was a man of destiny.
His wife, whom he had met rather late in her life,
Preferred to sleep, or complain of the food and the weather.
For their little garden was less than paradisiac,
With its artificial sun that either scorched or left you cold,
And savants with cameras eternally hanging around,
To perpetuate the only male quagga in the world.

Perhaps that was why he failed to do it himself.
It is all very well for goats and monkeys –
But the last male of a species is subject to peculiar pressures.
If ancient Satan had come slithering in, perhaps . . .
But instead the savants, with cameras and notebooks,
Writing sad stories of the decadence of quaggas.

And then one sultry afternoon he started raising Cain.
This angry young quagga kicked the bars and broke a camera;
He even tried to bite his astonished keeper.
He protested loud and clear against this and that,
Till the other animals became quite embarrassed
For he seemed to be calling them names.

Then he noticed his wife, awake with the noise,
And a curious feeling quivered round his belly.
He was Adam: there was Eve.
Galloping over to her, his head flung back,
He stumbled, and broke a leg, and had to be shot.[43]

Quaggas are the focus of two beautifully illustrated children's books. Tamara Green's *The Quagga* narrates the life and extinction of the animals.[44] The same themes are covered fictitiously in *The Quagga's Secret, Imfihlo yeQwahhashi*, by Hamilton Wende where the last surviving quagga longs to share his secret of happiness with Dai-Kwon, a Bushman, but is shot before he can do so.[45] The book is written in both English and Zulu and provides factual information about quaggas in an appendix. The quagga of King Louis XVI has a minor part in *All That Glitters*, the story of an orphan boy in the Royal Court at Versailles.[46] The science fiction work *Seven Moon Circus* features both a "Sea of Karoo" and a quagga, and shares with *All That Glitters* a person leaping onto a quagga and riding it – an extremely unlikely feat.[47] The novella *Born with the Dead* (1975) featured recreated quaggas and other extinct species and, of course, the extraction of quagga DNA from a hide provides the basis for Michael Crichton's novel *Jurassic Park*.[48]

Non-fiction accounts of quaggas include David Barnaby's books *Quaggas and Other Zebras* (1996) and *Quagga Quotations: a Quagga Bibliography* (2001), Reinier Spreen's *Monument voor de Quagga: Schlemiel van de uitgestorven Dieren* (2016); and an outstanding undergraduate thesis, Elizabeth Cary's *History and Extinction of the Quagga: Barking Horse of the Karroo* (1970).[49] In addition, two books about zebras provide good coverage of quaggas: Dorcas MacClintock's *A Natural History of Zebras* (1976) and *Zebra* by Plumb and Shaw (2018).[50] Other books and websites about extinct animals include information about quaggas. The historians Sandra Swart and Harriet Ritvo have also introduced quaggas to a new academic field: animal history.[51] The present book is a biologically oriented entry in that field.

Finally, the animated film *Khumba* (2013), produced by Triggerfish Animation Studios, follows the adventures of the eponymous zebra living in the drought-stricken Karoo. Khumba's quagga-like reduction in striping sets him apart from the rest of his herd and leads to his being ostracized – hence serving as a visual metaphor for issues of South Africa's troubled history of race. Khumba's journey to the waterhole, the site where zebras first received their stripes, brings encounters with animals that teach the value of diversity. The South African director, Anthony

Silverston, conceptualized the central theme of the film as learning to become comfortable with one's identity.[52]

There appears to be no literary work on quaggas between the 1834 publication of Pringle's *African Sketches* and "The Last of the Quaggas" in 1948.[53] But since the 1960s, quaggas have become subjects for visual artists, writers, and in popular culture where they are known for various reasons, including speaking deeply to matters of mortality and extinction. The poems and works of fiction cited in this section have all been published in the last few years. How extraordinary that an animal extinct for over a century and never well known outside of South Africa should now have assumed such a prominent role.

Museum Specimens

In museums, the hides and bones of quaggas shot in South Africa joined the earthly remains of quaggas that had lived in captivity.[54] Twenty-three mounted quagga specimens, a mounted head and neck, seven skeletons plus several skulls and a few assorted bones grace the collections of various institutions.[55] Most quagga artifacts are in Europe and the United States, although the Iziko South African Museum in Cape Town houses a female foal from Nelspoort in the Western Cape, the animal's skull and some of her foot bones, and the Ditsong National Museum of Natural History in Pretoria contains a skull. Over the years, specimens have been lost, including two skulls destroyed when the Royal College of Surgeons Museum in London was bombed during World War II, and a mounted specimen, the Königsberg quagga, was also obliterated during that war when drunken soldiers threw it out of a window.[56] More encouragingly, the quagga skeleton at the Grant Museum in London gained a missing left hind leg and a right scapula when computed tomography of the corresponding remaining bones was used to make replacements by 3D printing.[57]

Quaggas from zoos in Berlin, London, and Paris are spending their afterlives in the museums of those cities. Sometimes, hides and skeletons went to different locations: the mounted hide of the quagga mare that Haes and York photographed in the London Zoo is in the Royal Scottish Museum in Edinburgh, and her skeleton is in the Peabody Museum at Yale University.[58] The quagga stallion that shared this mare's enclosure in the London Zoo from 1858 until his untimely death in 1864 provided a skeleton for the British Museum and probably also the hide of the mounted specimen in Wiesbaden, Germany.[59]

Hides and skeletons were the focus of attention during the era when many biologists studied comparative morphology. Today, most are on public display in museums. There are mounted specimens in Austria (Vienna), France (Lyon and Paris), Italy (Milan and Turin), the Netherlands (Leiden; two specimens) Russia (Kazan), Sweden (Stockholm), Switzerland (Basel), and the United Kingdom (Edinburgh, London, and the Natural History Museum at Tring, which is part of the Natural History Museum, London).[60] Germany is rich in specimens, for besides skeletons and skulls at several locations, there are mounted specimens in Bamberg, Berlin, Darmstadt, Frankfurt am Main, Mainz, Munich, and Wiesbaden. Seven skeletons and a number of skulls (which possibly include equines other than quaggas) are in France (Paris), Germany (Berlin, Munich, Stuttgart, and Tubingen), Italy (Milan and Turin), South Africa (Pretoria), Switzerland (Basel), the Netherlands (Leiden), the United Kingdom (Bristol, Edinburgh, and London), and the United States (New Haven and Philadelphia).[61]

Considering these specimens brings us to Reinhold Rau of the South African Museum in Cape Town (now the Iziko South African Museum), who is frequently cited in this book. He was a taxidermist, a researcher of quaggas, and the engineer of their rebreeding. A colleague observed, "There were many strings to Reinhold's bow; he wasn't just a taxidermist," and this was indeed the case.[62] Described as "compassionate and generous," Rau mentored disadvantaged people at the South African Museum and during apartheid taught swimming to a racially mixed group of young people.[63] These activities led a colleague to write in his obituary, "So do not look for the soul of Reinhold Rau among the dead – he is not to be found there. Rather look for him among living things, whether these are people or tortoises or frogs."[64] To the category of living things also belong the zebras of the Quagga Project.

Reinhold Eugen Rau (1932–2006), who was born in Friedrichsdorf near Frankfurt am Main, Germany, wrote "From my early childhood I was fascinated by nature," and this sentiment is clear from his life.[65] Beginning in 1948, he trained at the Senckenberg Museum in Frankfurt before emigrating to South Africa in 1959 to work at the South African Museum in Cape Town, where he mounted animal skins, made models, and constructed dioramas. It was probably an uncomplicated transition for a skilled white man, from Germany to South Africa before South Africa's exit from the British Commonwealth. West Germany and South Africa were both allied with the United Kingdom, which was just then coming to terms with a changing situation in Africa. Against this political and societal backdrop, Rau's early work at the Museum involved its most controversial project: the Bushman dioramas prepared by James Drury. The dioramas recreated scenes of "bushmen" existence, with plaster casts of their living descendants providing lifelike models.[66] Dioramas with imitations of life forms made by casting is a classic form of museum representation, but this treatment of Bushmen put South Africa's indigenes in a category otherwise reserved for animals. This situation did not seem to strike museum officials, most white museum visitors, or Rau himself as anything unusual. On Drury's retirement, Rau became responsible for the molds and wrote an approving description of his technique.[67]

The Khoe-San were considered to have become extinct in South Africa.[68] Disproving and mitigating extinction was a thread in Rau's career, and examples abound: he discovered populations of the geometric tortoise, *Psammobates geometricus*, and helped establish a reserve for this critically endangered species; he also worked to preserve the Cape clawed frog, *Xenopus gilli*, another endangered species.[69] And then there were quaggas. Seeing the quagga foal in the South African Museum in Cape Town that had been inexpertly stuffed with clay and hemp, Rau vowed to do better and so in 1969 and 1970 he remounted the specimen and his lifelong quagga venture began.[70]

In the 1970s, Rau traveled to Europe and personally examined most of the twenty-three mounted specimens, providing careful descriptions, measurements, and photographs of them.[71] He noted that specimens in Berlin, Leiden, and Stockholm were stuffed with straw, and the Paris quagga still had the original wood mount from the eighteenth century. Modern taxidermy mounts the hide on a polyurethane form which ensures a good fit – something not always achieved with older mounts such as the hide of the Vienna quagga that Rau described as "obviously stretched."[72]

Rau found that some quagga hides were in good condition, but variously described others as cracked, patched, dirty, partly moth-eaten, partly destroyed by fire, shrunk, stretched, and crudely repaired.[73] The Munich quagga had several large cracks in its hide, but Rau restored the specimen so these were no longer apparent; he changed the stance of the animal, too – converting a forlorn-looking animal to one whose head is now proudly raised.[74] This specimen's hide had been obtained in 1834, mounted in Munich, and slightly damaged during World War II, but Rau restored it as an exhibit for a new natural history museum using donations from a Munich brewery.[75] Besides working restorative

Figure 7.1. Reinhold Rau and the quagga foal in the Iziko South African Museum, Cape Town, 1980. Photograph SAB 17167. Courtesy of the National Archives and Record Services of South Africa[76]

wonders on the Munich quagga and the foal in the South African Museum, Cape Town (Figure 7.1), Rau also remounted the three specimens at the Mainz Naturhistoriches Museum.

Rau noted that several specimens had faded, either over the entire animal or more locally, and he speculated that this was the cause of the unstriped faces of the Milan quagga and the Edinburgh quagga head.[77] Fading is presumably the reason that most quagga hides have a light brown or yellowish color, which is lighter than the chestnut color of quaggas in most paintings and of the "bright bay colour" described by Edwards for the type specimen; the dark brown hides of the Amsterdam, Berlin, and Paris quaggas are exceptional.[78] The biologist Richard Lydekker commented in 1904 on the fading that had already occurred in quagga hides:

I now believe, in fact, that the difference between the coloration of the stuffed Quaggas and the figures taken from living animals or fresh skins is entirely due to fading. On the head, neck, and fore-quarters the original blackish-brown stripes have faded to a brownish fawn similar to that of the hind-quarters; while the fawn intervals between the black stripes have bleached to white. The result of this is to produce a type of coloration quite distinct from the original one – namely, a fawn-coloured animal with white stripes. A similar effect is produced in the living animal, as exemplified in York's picture, by photography.[79]

Contemporary museum practice promotes specimen preservation by providing the appropriate temperature and humidity, by excluding pests, and by reducing damage from light, particularly ultraviolet light. But before these practices were instituted, hides were damaged when kept under less-than-ideal conditions, although the fact that DNA nevertheless survived has proved fortunate.

The presence of six quagga specimens within a radius of 50 kilometers (31 miles) from the center of Frankfurt am Main inspired me to visit Germany in the spring of 2018. These specimens are housed in museums in Darmstadt, Frankfurt am Main, Wiesbaden, and Mainz, the first three of which were built in the early years of the twentieth century, while the fourth is housed in a former monastery. All four buildings are stately structures that reflect civic pride and an appreciation of knowledge. Their collections are extensive and, whereas those of the Frankfurt and Mainz museums are confined to natural history, the Darmstadt and Wiesbaden museums also include art.

I began at the Senckenberg Museum in Frankfurt am Main with its attractive façade of light red stone flecked and splashed with cream coloration. Two petrified tree trunks estimated to be 250 million years old stand outside the museum and further fossils abound inside: a *Tyrannosaurus rex*, a mastodon, and a variety of specimens from the Messel Pit, a site not far from Frankfurt that is rich in fossils from the Eocene (36–57 million years ago). The quagga (said to be a stallion, but no sex organs are present) shared a display case with a mountain zebra which was considerably larger and whose black and white stripes contrasted with the faded brown colors of the quagga. I spent some time contemplating the two species, but it was noticeable that most other museum visitors passed them with only a glance.

The Hessisches Landesmuseum Darmstadt is an imposing building with a neoclassical main hall that leads to galleries devoted to archeology, art, culture, and natural history. A display spanning an entire gallery wall featured hundreds of specimens in a demonstration of animal biodiversity. Elsewhere in the gallery, in contrast, there were artifacts of extinction: an ivory-billed woodpecker, a thylacine (a carnivorous marsupial also known as the Tasmanian tiger), the skull of a Great Auk, the egg of

Figure 7.2. Quaggas and other zebras at the Mainz Naturhistoriches Museum. The stallion is behind the foal and the mare faces the camera. Photograph by the author. (A black and white version of this figure will appear in some formats. For the color version, please refer to the plate section.)

an Elephant Bird, and, of course, a quagga.[80] The juxtaposition of what still survives and what has so recently been lost made a powerful contrast. Appropriately for a museum that features a fine art collection, the 1820 painting of a quagga by Christian Wilhelm Karl Kehrer hung near the taxidermy specimen.

At the Mainz Naturhistoriches Museum, Dr. Carsten Renker, the Curator for the Zoological Collection, greeted me warmly and took me to the zebra exhibit, which included a mountain zebra, several subspecies of plains zebras, and quaggas. My photograph of the exhibit (Figure 7.2) shows the mare on the left facing the camera and the stallion in side view behind the foal. Behind the stallion is a specimen of Burchell's subspecies whose body stripes extend to the rump. A plains zebra with conspicuous shadow stripes stands at the back of the scene. The collection had been developed by a curator who wanted a complete representation of animals and so acquired specimens of a stallion, mare, and foal. Rau remounted the adults in 1980 and 1981 and made a life-sized model of the foal, incorporating remnants of hide from the original specimen that had been burned in a bombing raid. Shortly after the war, one of the quagga specimens was put up for sale, but the money offered was not substantial and no sale was made; this situation has changed, and the quaggas in the museum collection are now considered valuable.[81]

Drs. Fritz Geller-Grimm, Susanne Kridlo, and Hannes Lerp welcomed me to the Museum Wiesbaden and explained that their quagga specimen was not on public display but was in a storage area. This largest of the five stallion specimens in museums has an impressive tail that extends several inches below the hocks. Because of a long tear in its hide, it had been relegated to a storage area that was also occupied by a Cape lion with a magnificent mane – predator and prey both far from their native land.

A week before seeing the specimens in Germany, I had a private viewing of the quagga foal in the Iziko South African Museum of Cape Town. I had examined this specimen twenty years previously, but now it was in a gallery that was closed to the public while renovations were being made. Denise Hamerton, Curator of Terrestrial Vertebrates, led me through the darkened gallery to the quagga's display case and pressed a button that lit it momentarily. It was good to see the foal again, but it made me think about the disappointment of the farming family at Kamferskraal, Nelspoort, who had cared for the motherless animal on their farm – feeding her goat and sheep's milk – only to have her die after a week.[82] They were probably aware that quaggas were becoming rare and might have been attempting their conservation in the same way that later farmers protected Cape mountain zebras and bonteboks on their land before reserves for them were established.

Perhaps I saw too many taxidermy specimens within too short a time. Their glass eyes and faded hides with occasional cracks evoked Robert Frost's line, "what to make of a diminished thing."[83] But I consoled myself with the thought that quaggas do survive in other ways as DNA, rock art, paintings, photographs, in the accounts of those who saw them, in the creative work of those who contemplate them, and phenotypically as the animals of the Quagga Project.

Molecular Biology Resolves Taxonomy

Until recently, opinions were divided on whether quaggas and plains zebras were the same, or different, species. The question "What is a species?" defies a simple answer. Even the author of *The Origin of Species* fretted that there was not "a clear & simple definition of what is meant by a species," and observed, "It seems to me that the term species is one arbitrarily given for convenience sake to a set of individuals closely like each other; &, that it is not essentially different from the term variety, which is given to less distinct & more fluctuating forms."[84] As noted by Darwin, variations exist within a species and sometimes these are so pronounced that biologists have described subspecies or varieties within the species. If the differences between two zebras are more substantial, they could be classified as different species. John Gray believed this to be the case when he described the plains zebra by the binomial name *Asinus burchellii* as a species distinct from quaggas, which had been described earlier.[85]

In contrast, the game hunter Frederick Selous believed that, "The true quagga was nothing but the dullest coloured and most southerly form of Burchell's zebra."[86] He supported his argument by describing the cline in coat appearance from well-striped plains zebras in Central Africa to quaggas at the most southerly part of the species' range.[87] Later authors made similar observations, and the cline is vividly illustrated by the photographs of plains zebras in Figure 3.1.[88]

There are different definitions of species but a commonly accepted one, the "Biological Species Concept," regards species as populations of similar individuals which breed with each other and produce fertile offspring, but do not usually breed with other species under natural conditions.[89] There were no records of quaggas breeding with plains zebras where these animals ranged together (or even when they were kept together in captivity), but absence of evidence is not evidence of absence, as the behavior of these animals was not studied in detail.[90]

Lacking records of interbreeding, the relatedness of different organisms can be studied by comparing their morphology. For example, although minor variations occur within a population of a species, observations of a wider range of differences between two populations might support the argument that they are different species. The subjectivity of this approach is obvious and could result in one taxonomist declaring that differences between two populations were substantial enough that they should be considered different species, and another viewing these differences as the variation that occurs within a single species. So established are these types of taxonomic approaches that their practitioners are termed "splitters" and "lumpers," respectively.

Richard Lydekker employed a comparative anatomy approach early in the twentieth century. He examined the skulls of two quaggas and several plains zebras: the former had a shallow, almost circular, depression of the skull, the "preorbital pit" or "face-pit," which was absent in the latter. He observed, too, that the pattern of facial stripes differed between quaggas and plains zebras. Based on these criteria, he declared that quaggas and plains zebras were separate species.[91] These appear to be minor differences on which to separate species and the preorbital pit, moreover, proved to be a variable feature: Reginald Pocock acknowledged its presence in the two skulls cited by Lydekker but observed that it was absent in two other quagga skulls that he examined and, additionally, was present in a plains zebra's skull.[92]

Given that the pre-orbital pit was of no use in taxonomy, other anatomical criteria were needed. Bengt Lundholm's study of dentition in nineteen plains zebras and eight quaggas making use of more characteristics and a larger sample size led him to the conclusion that plains zebras and quaggas were separate species.[93] Applying the same diagnostic criteria to the dentition of a skull in the Transvaal Museum (now the Ditsong National Museum of Natural History) labeled "zebra, no records," Lundholm revealed that it came from a quagga. This specimen (catalog number TM 10161) is regarded as being one of the few quagga artifacts in South Africa.[94]

Klein and Cruz-Uribe measured 23 cranial dimensions from four skulls that were well documented to have belonged to quaggas and compared this data with measurements from 164 skulls of plains zebras and 63 skulls of mountain zebras, as well as skulls from other *Equus* species. This investigation demonstrated that quaggas and plains zebras differed from each other to approximately the same extent as mountain zebras and plains zebras differed from each other.[95] Other scientists, though, when comparing cranial dimensions, found that plains zebras were more similar to quaggas than were mountain zebras.[96] Comparative studies of dentition were no more conclusive: a report in 1988 suggesting that quaggas may not have been a separate species from plains zebras was followed nine years later by data from the same source showing a greater similarity between plains zebras and mountain zebras than either of these zebras had with quaggas.[97]

While comparative anatomy can be a powerful tool in biology, in this instance it failed to clarify whether quaggas and plains zebras were the same species or not. Fortunately, different techniques based on analysis of DNA and proteins can address taxonomic questions and had, in fact, already done so. The morphological investigations published between 1988 and 2004 came after molecular studies that resolved quagga taxonomy for many biologists.

Molecular studies on quaggas originated from Rau's inventiveness, serving as a fine example of Louis Pasteur's aphorism that "Chance favors only the prepared mind." While remounting the Cape Town foal in 1969 and 1970 and the Mainz quaggas in 1980 and 1981, Rau saved muscle and connective tissue that had been left attached to the salt-treated hides. Taxidermists usually remove such tissues from hides during specimen preparation, but fortunately the original taxidermists did not work to such standards. Rau saw an opportunity: perhaps analysis of these quagga tissues could determine whether quaggas and plains zebras were the same or different species? In 1979, he contacted the zoologist Dr. William F. H. Ansell to enquire about the feasibility of tests on quagga tissue. Ansell, who was knowledgeable about zebras, replied that "there does not seem to be any meaningful cytological tests that could be carried out on such long dead material."[98] This statement was a reasonable one: salt-treated tissue that has dried and aged does not seem to hold much promise for chemical analysis; however, this assessment was not the end of the matter. Two years later, Oliver Ryder of the San Diego Zoo sent Rau a request for zebra tissues to test, and he responded with muscle and connective tissue that had been attached to the hide of a quagga in the Mainz museum. This tissue went to two groups of scientists in California, with Ryder himself collaborating with Jerold M. Lowenstein at the University of California, San Francisco, to demonstrate that quagga proteins had a closer resemblance to those of plains zebras than to either mountain zebras or Grévy's zebras.[99]

Exciting as this discovery was, it paled in comparison with the news that Russell Higuchi and his team at the University of California, Berkeley, were able to extract DNA from the quagga muscle and connective tissue.[100] Both nuclear and mitochondrial DNA were present, but they identified two samples as DNA from mitochondria, the cell organelles that supply most of the energy used by aerobic organisms. Mitochondrial DNA is an attractive molecule for studying evolution because it mutates at a much faster rate than nuclear DNA. Over time, mutations occur and substitute one nucleotide for another, and these changes in nucleotide sequence can be detected by DNA sequencing.

Much of the quagga DNA had been degraded, and the remainder had broken into shorter pieces; nonetheless, Higuchi and his colleagues determined the sequence of 229 nucleotides in two short lengths of mitochondrial DNA. Comparison of this DNA with the corresponding nucleotide sequences from a plains zebra revealed a difference at only two nucleotides. The similarity in the DNA from quaggas and plains

Figure 1.2. A Hartmann's mountain zebra and a plains zebra at left photographed in Etosha National Park, Namibia. Photograph by Yathin S Krishnappa. Creative Commons Attribution-Share Alike 4.0 International license. (A black and white version of this figure will appear in some formats.)

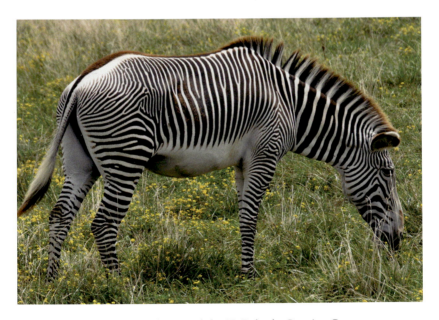

Figure 1.3. Grevy's Zebra. Photograph by T. Bobosh. Creative Commons Attribution-Share Alike 2.0 Generic license. (A black and white version of this figure will appear in some formats.)

Figure 2.2. Petroglyph of a quagga incised on andesite rock, 1000–2000 BP. Image id: 01612928374. Courtesy of the British Museum. (A black and white version of this figure will appear in some formats.)

Figure 2.3. Een jonge kwagga. Female quagga foal depicted by Robert Gordon in 1777. Gordon, The Gordon African Collection, (RP-T-1914-17-190). Courtesy of the Rijksmuseum. (A black and white version of this figure will appear in some formats.)

Figure 2.4. Zebra femina. George Edwards painted this quagga mare in 1751. Plate 223 in Edwards, Gleanings of Natural History, 1758. (A black and white version of this figure will appear in some formats.)

Figure 2.5. The Quahkah. Plate 15 in Daniell, African Scenery and Animals, 1804–1805. Courtesy of the British Museum. (A black and white version of this figure will appear in some formats.)

Figure 2.6. Een Quagga. Aert Schouman painted this watercolor of a quagga stallion in 1780. Courtesy of the Teylers Museum. (A black and white version of this figure will appear in some formats.)

Figure 2.7. Le Couagga. Nicolas Maréchal painted the quagga stallion of King Louis XVI in 1793. https://en.wikipedia.org/wiki/Quagga#/media/File:Quagga.jpg. (A black and white version of this figure will appear in some formats.)

Figure 2.8. Quagga, Asinus quagga. Waterhouse Hawkins painted these quaggas at the Knowsley Menagerie of Lord Derby. Plate 54 in Gray et al., Gleanings from Knowsley Hall, 1850. (A black and white version of this figure will appear in some formats.)

Figure 2.9. A Male Quagga from Africa, the First Sire. Jacques–Laurent Agasse painted Lord Morton's stallion in 1821. Object number RCSSC/P 278. Courtesy of the Royal College of Surgeons. (A black and white version of this figure will appear in some formats.)

Figure 2.10. Equus quagga. The Quagga. Detail of Plate 2 in Harris, Portraits of the Game, 1840. (A black and white version of this figure will appear in some formats.)

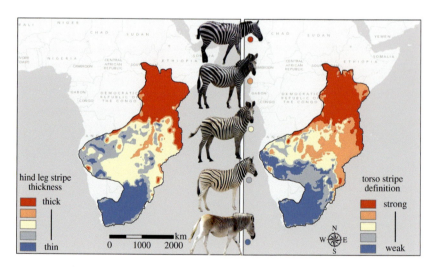

Figure 3.1. "Predicted levels of hind leg stripe thickness (left) and torso stripe definition (right)." From figure 2 in Larison et al, "How the zebra got its stripes: a problem with too many solutions," Royal Society Open Science January 1, 2015. Attribution–Share Alike 4.0 International license. https://doi.org/10.1098/rsos .140452. (A black and white version of this figure will appear in some formats.)

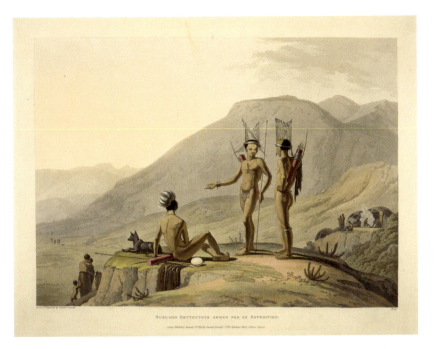

Figure 4.1. Bush-men [Khoekhoe] armed for an expedition. Plate 2 in Daniell, African Scenery and Animals, 1804–1805. Courtesy of the British Museum. (A black and white version of this figure will appear in some formats.)

Figure 4.3. Boors returning from hunting. Plate 11 in Daniell, African Scenery and Animals, 1804–1805. Courtesy of the British Museum. (A black and white version of this figure will appear in some formats.)

Figure 7.2. Quaggas and other zebras at the Mainz Naturhistoriches Museum. The stallion is behind the foal and the mare faces the camera. Photograph by the author. (A black and white version of this figure will appear in some formats.)

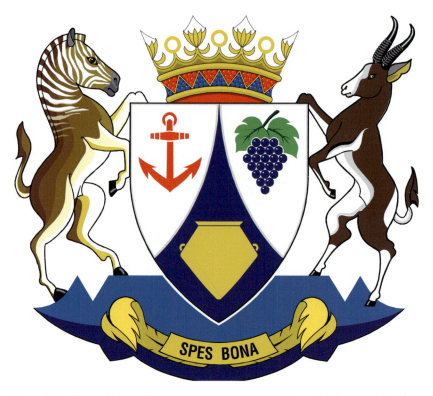

Figure 7.3. Coat of Arms of the Western Cape. Designed by Frederick Brownell and adopted by the Western Cape Provincial Parliament in 1998. (A black and white version of this figure will appear in some formats.)

zebras argues for a close relationship between them, and it is even possible that their DNA might be identical as postmortem changes in quagga DNA could account for the two nucleotides that differed.[101]

The rate of nucleotide change in DNA can serve as a molecular clock and so the time needed to produce the zero or two difference in nucleotides between corresponding DNA sequences of quaggas and plains zebras was clearly less than the time needed to give the twelve nucleotide difference that occurred between corresponding DNA sequences of mountain zebras and quaggas, and the fourteen nucleotide difference that existed between corresponding DNA sequences of quagga DNA and horse DNA.[102] Collectively, these data argue that the evolutionary lines leading to horses and zebras diverged, and later the evolutionary line leading to mountain zebras diverged from the one leading to plains zebras.

The analysis by Higuchi and his team was technically remarkable because it involved screening 25,000 samples in order to obtain the two samples of mitochondrial DNA. Additionally, this investigation was carried out before the innovation of the polymerase chain reaction made it possible to make multiple copies of DNA molecules and before fast sequencing techniques had been developed.

Not all biologists were convinced that the 229 nucleotides sequenced were representative of the entire mitochondrial genome.[103] However, subsequent studies of longer sequences extracted from thirteen quagga specimens indicated that the nucleotide sequences of mitochondrial DNA were similar in quaggas and plains zebras, confirming that these animals belong to the same species.[104] This data also showed that quaggas lacked genetic diversity and had diverged from other plains zebras 120,000 to 290,000 years ago.[105]

With rapid improvements in sequencing techniques, information on quagga genomes expanded from just 229 nucleotides of mitochondrial DNA in 1984 to the entire mitochondrial genome in 2013 and to the whole nuclear genome comprising billions of nucleotides in 2014.[106] In this latter study, Jónsson et al. compared complete genomes of asses, zebras, and a quagga specimen. They showed that the genus *Equus* arose in the New World, migrated into the Old World 2.1 to 3.4 million years ago, and diverged between 1.28 to 1.59 million years ago into the evolutionary lines that gave rise to plains zebras, mountain zebras, and Grévy's zebras.[107] These genomic studies estimated that quaggas diverged from other plains zebras 233,000 to 356,000 years ago and established that there had been positive selection for two genes in quaggas, including the *SEMA5A* gene, which in humans is involved with autism and control of hippocampal volume – a discovery that raises a question of whether there were behavioral differences between quaggas and other plains zebras.[108] Jónsson et al. showed that there had been positive selection for two different genes in *Equus quagga boehmi*, including the *SLC9A4* gene that codes for a protein that protects the animals in unfavorable environments.

The most recent investigation of the complete genomes of plains zebras by Casper-Emil Pedersen and his colleagues shows that quaggas diverged from other plains zebras about 367,000 years ago, a period that includes at least two Ice Ages.[109] As noted in Chapter 3, changes in temperature – and probably aridity, too – during Ice Ages and their interglacial periods would have affected the habitats of quaggas, which in turn probably influenced evolution of their coat characteristics. Alternatively, if genetic drift were a factor in the evolution of quaggas, hundreds of thousands of years would have provided time enough for this process to occur in an isolated population.

Pedersen et al. showed that plains zebras possibly originated in the Zambezi-Okavango wetland areas (in present-day Botswana and Zambia). Migration north to equatorial Africa and south to southern Africa resulted in nine extant populations that do not always correspond with subspecies based on morphology. For example, *Equus quagga boehmi* and *Equus quagga crawshayi* both comprise two or more populations that differ genetically. This information is important, as it can identify genetically distinct populations that should be conserved. Similar studies have shown that what was understood to be a single species of Galapagos tortoise was two distinct species and that giraffes comprise four distinct species – two of which contain just a few thousand individuals and which should, therefore, be a conservation priority.[110]

Although some plains zebra subspecies have proved to be genetically diverse, quaggas and the Namibian plains zebras are populations that are both distinct genetically and correspond with two subspecies recognized by taxonomists: quaggas and Burchell's subspecies (*Equus quagga quagga* and *Equus quagga burchellii*).[111] This research also addressed whether quaggas differed significantly in their genetics from other plains zebras and found that they had diverged less from nearby populations than had the population in northeastern Uganda. Genetic differences between populations of plains zebras, together with evidence for selection for different genes in quaggas and *Equus quagga boehmi*, will be examined in Chapter 9 when considering whether Quagga Project zebras are, in fact, quaggas.

Many biologists were surprised to learn that tissues taken from animals that had been dead for more than 100 years could provide DNA – now termed aDNA, or ancient DNA – that could be sequenced.[112] This work served as proof of concept that the DNA of extinct organisms could be studied and has launched the discipline of "paleogenomics" or "molecular archeology."[113] And just as Lord Morton's quagga provided the (flawed) scientific basis for the writings of August Strindberg and Émile Zola, so the survival of intact DNA in quagga tissues helped Michael Crichton to envision the persistence of dinosaur DNA in their blood cells preserved in amber-embedded blood-sucking insects for his 1990 novel, *Jurassic Park*.[114]

Molecular studies indicating that quaggas belonged to the same species as plains zebras brought about changes in classification. Because of the rules of scientific nomenclature, the new development was not that quaggas were now plains zebras, but that all these animals are now named "quagga." Both Boddaert and Gmelin had proposed the binomial name *Equus quagga* and had described this new species before Gray named the plains zebra *Asinus burchellii*, later renamed *Equus burchellii*. Under the practice of the *International Code of Zoological Nomenclature*, the oldest valid name of a species has precedence, and so all plains zebras are now *Equus quagga*.[115] Changes in names are often vexing to non-taxonomists – as is clear to dinosaur enthusiasts who have had to switch between the names *Brontosaurus* and *Apatosaurus*. Nonetheless, the name *Equus quagga* reflects a refinement in our understanding of plains zebras, and the least-striped subspecies of southern African zebras can be distinguished by the tri-nomial name, *Equus quagga quagga*. It took a long time to reach this conclusion. Debate about whether quaggas and plains zebras were con-specific led to research in comparative morphology and yet more dis-agreement. Now the matter is settled and, although a name change seems a relatively minor matter, this new classification has had profound impli-cations for the feasibility of the Quagga Project, the subject of Chapter 8.

Quaggas in the New South Africa

In the post-apartheid era, the country needed new, more inclusive ways to represent itself as unified.[116] The South African Museum, Reinhold Rau's employer, took on the challenge of transformation by renaming itself Iziko, the Xhosa word for hearth, to signify the space as a hub for cultural interchange. The museum also closed the controversial Bushman diorama in 2001. Public opinion, not least among the descendants of the

Khoe-San, held that a natural history museum in the New South Africa should not represent the indigenous people in this way. Reinhold Rau had participated in training programs that introduced disadvantaged youth, perhaps including the descendants of the Khoe-San, to museum studies, but from retirement he objected to the shuttering of the diorama.[117]

Extinction, race, and a new politics of representation met in the Bushman diorama. These politics were in play around quaggas as well. The reclassification of plains zebras – joining together into a single entity, *Equus quagga*, that which had been separated by coat coloration – occurred at about the same time that South Africa abandoned a color-based classification of its citizens. The general public did not follow the taxonomic debates about zebras, but culture brokers were looking for

Figure 7.3. Coat of Arms of the Western Cape. Designed by Frederick Brownell and adopted by the Western Cape Provincial Parliament in 1998.[118] (A black and white version of this figure will appear in some formats. For the color version, please refer to the plate section.)

new symbols. So it was that Michel Clasquin, the editor of *Quagga Kultuur: Reflections on South African Popular Culture* viewed the appearance of quaggas as symbolic of South African culture: the dark and light stripes of their faces, necks and forequarters representing racially different cultures but shading into each other posteriorly to give a blended, authentic South African culture. Clasquin regarded the restoration of quaggas as representing the resurgence of South African culture; however, when pondering selection based on coat coloration, he enquired, "Is it still, in South Africa, important that everything be the right colour?"[119]

This chapter concludes with an official instance of the New South Africa embracing quaggas. The coat of arms of the province of the Western Cape (Figure 7.3), designed by F. G. Brownell and adopted in 1998, features a quagga and a bontebok, with symbols representing the wildlife and topography of the province (the two animals stand on a stylized depiction of Table Mountain), its maritime and agricultural resources, and its Khoekhoe heritage (the clay pot at the base of the shield).[120] Like quaggas, bonteboks, *Damaliscus pygargus*, were endemic to the Cape, but they narrowly escaped the fate of quaggas. The choice of these two species on the provincial coat of arms provides both an unvarnished representation of provincial history and a provocative commentary on the motto "Spes Bona" (Good Hope).

8 · Rebreeding

The quagga became extinct through man's ignorance and greed. It wasn't a natural occurrence. It is our moral duty to rectify that mistake.

Reinhold Rau[1]

Rau's recognition of a "moral duty" to right a wrong by rebreeding quaggas gained support from many people and organizations: hunters, who also recognized an obligation; South Africans, who viewed rebreeding as a matter of national prestige; landowners, who were willing to accommodate zebras; and many donors, who were undoubtedly stirred by the prospect that animals resembling quaggas might live in their former habitats.[2] Rau's case for rebreeding was supported by DNA studies showing that quaggas were not a separate species, but were a subspecies of *Equus quagga*, and so he argued that the alleles for their distinctive coat coloration should still be present in the least-striped and most brown plains zebras and could be retrieved. Selective breeding over thirty years for plains zebras with fewer stripes and browner coloration has produced animals termed "Rau quaggas."

Prospects of Rebreeding

Sequencing of genomes from extinct animals has provided data for scientists and plot possibilities for writers of science fiction, but long before recombinant DNA technology made genetic engineering possible, selective breeding to restore quaggas was first suggested by James Ewart, the same biologist whose breeding experiments between a zebra and domestic horses had disproved telegony. Writing on May 28, 1900, to Mr. Tegetmeier, editor of *The Field*, Ewart described a plains zebra that had distinct stripes on its neck but was otherwise almost white. Ewart speculated, "Were a pair of white zebras isolated & protected [bred only with each other] a new quagga might be produced."[3]

In the mid-twentieth century, Heinz Heck, Director of the Hellabrunn Zoo in Munich, and his brother, Lutz Heck, Director of the Berlin Zoo, attempted to rebreed extinct animals called aurochs, the wild ancestors of cattle. The Heck brothers had separate projects in which they bred together cattle that they judged to have auroch-like characteristics, and then selected their most auroch-like offspring.[4] Reichsmarschall Hermann Göring encouraged these breeding experiments to stock German-occupied lands in Eastern Europe. The descendants of these animals, known as "Heck cattle," still exist; they are not aurochs but have some of their characteristics, such as large size, long horns, and fierce dispositions.[5]

The Heck brothers also attempted to rebreed tarpans, the extinct ancestors of modern horses, by breeding Przewalski's horses with various equine breeds thought to be similar to tarpans. The resulting animals have been termed "Heck horses."[6] And, in what can be viewed as the first attempt to rebreed quaggas, Heinz Heck mated poorly striped plains zebras to produce a foal whose hindquarters reportedly lacked stripes. However, this effort halted with the loss of the breeding stock during the war.[7]

No doubt Lutz Heck recalled his brother's breeding experiment with poorly striped plains zebras when visiting Etosha National Park in South-West Africa during the early 1950s. Observing the diversity in striping and brown coloration within a herd of plains zebras, he saw occasional animals with fewer stripes and browner color and speculated that selective breeding between such animals could produce a quagga-like animal.[8] The same conclusion was reached later by Reinhold Rau, who had carefully examined most of the mounted specimens of quaggas and had concluded – like Pocock and Selous before him – that quaggas were not a separate species but were the least-striped and most brown form of plains zebras.[9] Rau had seen one of Heck's cattle when he was a boy and was intrigued by the notion of rebreeding extinct animals. Now he saw the possibility of producing quagga-like animals from plains zebras.

During the 1970s and early 1980s, Rau discussed his ideas for rebreeding quaggas with, and received the support of, several South African biologists including Drs. Banie L. Penzhorn and G. L. ("Butch") Smuts (authorities on zebras), and Dr. Uys de V. ("Tol") Pienaar (a zoologist). The latter wrote enthusiastically: "we should be successful in breeding back a true likeness of the extinct quagga in a relatively short space of time."[10]

At this early stage, Rau even identified a donor willing to provide space for a breeding population of plains zebras at his private game park. His request for institutional support, however, proved unsuccessful.

In 1975, the Natal Parks Board viewed, "the proposal to breed a quagga-like animal to be merely an academic exercise of very dubious conservation value."[11] Later, Dr. Pienaar, serving as an advocate for Rau's proposals, fared no better and ruefully reported, "my suggestion to the Parks Board for a project to selectively breed plains zebras to produce an animal resembling the extinct quagga met with flat and complete opposition."[12] Rau persisted and in 1982 he made proposals for rebreeding quaggas to the National Zoological Gardens of South Africa and the Endangered Wildlife Trust, but both organizations declined support, citing "a very tight budget," and "limited funds," respectively.[13]

Adding to the practical and institutional hurdles, the question remained of whether plains zebras could provide suitable stock for a rebreeding effort. An early proposal to rebreed quaggas had been criticized on the basis that they were a different species from other zebras and so their unique genetic material was lost with their extinction.[14] In 1984, Rau received the heartening news that this genetic barrier had fallen; short lengths of mitochondrial DNA from quaggas and plains zebras were identical, or almost so, which showed that the animals were conspecific – an essential precondition for the selective breeding of animals resembling quaggas from plains zebras.[15]

Even though rebreeding now had a justification, the question remained of whether it was worthwhile: in June 1985, the authorities in Namibia rejected Rau's request for plains zebras with the uncompromising statement, "This Directorate regards your proposed project on the extinct quagga as academic and of little importance to conservation ... we regret to inform you that we cannot make any Burchell's zebra available to you."[16] This evaluation echoed the criticisms of the Natal Parks Board a decade earlier. Whether rebreeding was a significant conservation goal seems to have been central to these funding decisions; this important question will be discussed in Chapter 9.

September 1985 represented a turning point in the form of a scientific publication and a new influential supporter. The journal *Experientia* published a paper showing that proteins in quagga tissues were much more similar to those of plains zebras than those of mountain zebras and Grévy's zebras, a conclusion that complemented the results from the sequencing of mitochondrial DNA.[17] At the same time, Dr. J. F. Warning, a retired veterinarian with experience in horse breeding who had recently contacted Rau, arranged a meeting for October 1985 at which Cape Nature Conservation (now called CapeNature) agreed to support restoration of quaggas – including providing accommodation and forage for zebras at the Vrolijkheid Nature Reserve near the Western Cape town of Robertson.

In December 1985, the Cape Department of Nature and Environmental Conservation offered further material assistance in the form of fencing and feed for the animals at Vrolijkheid and agreed to negotiate with the authorities in Namibia for zebras. Conversations with personnel from the South African Museum then followed and, on March 20, 1986, the Director of the South African Museum, Dr. M. A. Cluver, decided that the "Quagga Project" would be part of its operation under the administration of a "Quagga Experimental Breeding Committee."[18] "Vrolijkheid," the name of the farm where the zebras would be housed, is Afrikaans for "Joyfulness," which surely must have epitomized the mood of Rau and his supporters with these changes of fortune.

The Quagga Project

Before any plains zebras could be caught, practical matters required attention, such as preparing an enclosure of 6,400 square meters at Vrolijkheid. In what was to be the first of many donations to the Quagga Project, a company heavily discounted the wire used for the fencing. In April 1986, Rau visited Etosha National Park to discuss with the game capture team from the Department of Nature Conservation and Tourism of Namibia the characteristics of plains zebras that should be selected for the Project, and later sent them photographs of suitable animals.

Despite these preparations, the capture of plains zebras in March 1987 was fraught, as Rau and the capture team had different ideas about which animals they should select. Rau, who was said to have "exacting standards" was as demanding in his selection of plains zebra as in his painstaking restoration of quagga taxidermy specimens.[19] He was disappointed by the "less suitable" coat coloration of four zebras that had already been captured before his arrival at Etosha National Park; in turn, "The capture team became disenchanted with me not liking most of the animals which were pointed out to me as 'good'."[20] Things were no easier for the animals; several died after capture, but at the end of a quarantine period, nine plains zebras arrived at Vrolijkheid on April 21, 1987. The South African Nature Foundation funded purchase of the animals, their transport, and the initial expenses of constructing their enclosure.

Four additional plains zebras from Natal (present-day KwaZulu-Natal) were brought to Vrolijkheid in November 1988; they were joined in August 1993 by Howie, a mare with good coat coloration from the

Umgeni Valley Reserve in Natal.[21] Although these animals belonged to the same subspecies, those from South Africa had less striping, whereas those from Etosha had browner skins.[22] In 1992 to 1994, the animals were moved to sites closer to Cape Town, including land on the edge of Table Mountain National Park, part of the Rhodes Groote Schuur Estate.[23] This move provided natural grazing for the animals that previously had required fodder and also put the Quagga Project zebras in the public eye from De Waal Drive.

Following the birth of the first foal in December 1988, the population of Quagga Project zebras steadily increased. In March 1998, eleven animals were released into the Karoo National Park, followed by three more in September 1998; later, Quagga Project zebras were released into the Addo Elephant National Park.[24] The Quagga Project increased its breeding stock in September 1999 by adding eight plains zebras with reduced striping from the Umgeni Valley Reserve in KwaZulu-Natal.[25] In return, the Umgeni Reserve gained eight zebras, seven blue wildebeests, seven kudus, and seven red hartebeests. This swap of twenty-nine animals for eight zebras indicated the value accorded to suitable breeding stock.

The important steps of releasing Quagga Project zebras into parks and reserves and obtaining appropriate animals for further breeding by the Quagga Project were formalized by the signing in June 2000 of an agreement between the Quagga Project Association and the South African National Parks (SANParks). In this cooperation, SANParks provided plains zebras with appropriate coat coloration from national parks to the Quagga Project, along with financial support for their transport. In return, the Quagga Project donated their animals to the Addo Elephant National Park and the Karoo National Park.[26]

Capturing zebras and moving them to a new location is challenging because darted animals dash in panic until the sedative takes effect. As noted in this chapter, there were several fatalities during the 1987 capture in Namibia, and five died during or shortly after the capture of sixteen animals in 2016. The Quagga Project has given the welfare of its animals the highest priority, and the latter incident led to the appropriate decision to use a different capture team.[27] By contrast, only two animals died among the over fifty that were moved in 2017–2018.[28]

Now that the Quagga Project is more established, animals will only be moved to improve the breeding stock at a particular location or if they are sold. Just keeping plains zebras together can also lead to fatalities, as happened when four young animals were killed by stallions bothering the breeding herds at Wedderwill.[29] There were similar instances of young foals being intentionally killed by stallions introduced to new herds during the establishment of Core Herds in 2017 and 2018. This loss is now prevented by waiting until the foals are at least three months old before they have contact with the new stallions.[30]

The selective breeding employed by the Quagga Project followed the same approach that humans have used for millennia to give rise to the different varieties of domesticated animals and cultivated plants that exist today – including the auroch-like and tarpan-like animals bred by the Heck brothers. Charles Darwin described the principle behind selective breeding: "Methodical selection is that which guides a man who systematically endeavours to modify a breed according to some predetermined standard."[31] The "predetermined standard" for rebreeding of quaggas is detailed in the "Quagga Project Management/Action Plan," which calls for "selectively breeding zebra with quagga-like characteristics" which are defined as, "Decreased body stripes; Body stripes not extending to the ventral midline; A chestnut basic colour on unstriped, upper parts of the body; Unstriped legs; Unstriped tail; Reddish muzzle."[32]

The visible characteristics of an organism such as the reduced numbers of stripes on quaggas' bodies and the chestnut color of their upper bodies are its phenotype, which is determined by both the organism's genes (its genotype) and its environment. Genes exist in alternative forms termed "alleles." For example, in Gregor Mendel's well-known experiments on inheritance in peas, one allele coded for a green seed coat, whereas a different allele of the same gene coded for a yellow seed coat.[33] Unlike pea coloration, where the two copies of a single gene – one from each parent – determine seed coat color, many characteristics are under the control of several genes, each of which may have more than one allele. Height, eye color, and hair color in humans are all examples of multigenic traits, and it is probable that coat coloration in quaggas is also under the control of multiple genes and alleles.

If the alleles determining coat coloration in quaggas are still present in populations of plains zebras, it ought to be possible to breed for them, "By bringing selected individuals together, and so concentrating the Quagga genes."[34] Applying this principle, the nineteen founder animals of the Quagga Project and those subsequently added were chosen for their coat characteristics, and their offspring were likewise selected, with unsuitable animals removed from the breeding stock.[35]

At each of the locations where Quagga Project zebras live, stallions and mares are free to choose their own mates; there has been no forced mating. Consequently, the selection of which animals are kept together is of prime importance. The Quagga Project has kept careful records, creating a studbook that allows the breeders to decide which animals will be present at each location and so avoid excessive inbreeding that might cause adverse results.[36] Professor Eric H. Harley, a geneticist from the University of Cape Town, who has overseen the genetic aspects of the Quagga Project since its inception, stated in 2019 that there have been no deleterious results of inbreeding among the Quagga Project zebras.[37]

Selection in each generation of those zebras most resembling quaggas in coat characteristics and then breeding from them has produced fourth- and fifth-generation progeny which have the same degree of striping on their faces, necks and forequarters as the founder population but significantly fewer stripes on their legs and hindquarters. Graphs on the Quagga Project website show the stripe reduction from the founder population to the fourth generation. The average number of stripes on the hind-body was reduced from 4 to approximately 2.5; stripes on the forelegs were reduced from just over 4 to less than 2, and there was a corresponding reduction from just over 8 stripes to 4 stripes on the hind legs. These data are *average* numbers of stripes for each generation, but there are now individuals lacking stripes on their forelegs and hindquarters.[38]

By July 2004, there were eighty-three Quagga Project zebras at eleven locations.[39] In November 2006, the Quagga Project reported selling forty-nine zebras for R290,500, leaving sixty-six animals.[40] Their numbers increased and in March 2017 there were 142 animals; by this time, 1,874 foals had been born into the Quagga Project and so, besides deaths, clearly there had been exacting selection, with undesirable zebras being sold to provide funds.[41] One that was sold later proved to be "an excellent quagga type animal," and unsuccessful attempts were made to buy it back.[42] Other zebras that originated at the Quagga Project are on private land, including those at the Wedderwill Game Reserve and the Bontebok Ridge Reserve.[43]

Rau Quaggas

The Quagga Project distinguishes those of its zebra with the best coat coloration as "Rau quaggas" – a term that differentiates the animals from extinct quaggas and celebrates Rau's pioneering contribution:

From our point of view they are "real" Quaggas (Equus quagga quagga), but since there has always been the possibility that there might have been other features of the original Quagga (Equus quagga quagga) that we have not selected for (because we do not know what those features, if any, might have been), we have chosen the term "Rau Quagga" to describe our recovered phenotype.[44]

Rau quaggas show the efficacy of selective breeding; however, their coat coloration differs in several respects from that of museum specimens and historical illustrations – having generally more stripes (including on the hind legs) and a lighter brown background color. The Quagga Project allows stripes on the hind legs by citing the authority in matters of selection and species: "Darwin recorded occasional stripes on the hocks in some Quaggas (Equus quagga quagga)." [45] In fact, Darwin described just one such animal, and this observation of leg stripes in quaggas is not corroborated by museum specimens and historical descriptions and illustrations, including the type specimen illustrated by George Edwards; perhaps the animal described as a quagga was a Burchell's subspecies. [46]

The goal of "chestnut basic colour on unstriped, upper parts of the body" has been harder to achieve. Commenting on changes that occurred in the first four generations of selected animals, the Quagga Project website admits that "Background colour estimates have so far shown no progressive change over time," and another evaluation of the Project observed that although there had been a reduction in stripes, the restoration of a brown coat color had been less successful. [47]

Some of the latest Quagga Project zebras have "an increase in the background brown colour . . . as if the pigment in the stripes has smeared out to give a uniform brown colour." [48] This development is encouraging as a brown or chestnut color is prominent in descriptions of historical quaggas: Boddaert described the color of the upper body as "fusco" (brown), and Gmelin used the word "castaneus" (chestnut). [49] The chestnut color is evident in many paintings of quaggas, including the type specimen with its "bright-bay colour." [50] Somerville, who had seen many live quaggas, described the color of the upper part of the body as "dark, glossy brown," and Harris, who had shot many of the animals wrote, "Colour of the neck and upper parts of the body, dark rufous brown, becoming gradually more fulvous [orange-brown]." [51]

It was estimated in 2020 that approximately twelve animals would qualify as Rau quaggas. [52] To achieve browner animals, Eric Harley and his colleagues have promised that, "animals with the darkest background colouration will claim priority in the ranking of Rau quaggas." [53] Similarly, Bernard Wooding of the Quagga Project acknowledged, "We've learned how to get rid of stripes; now we have to work on the brown color." [54] And the Chairman's Report from the June 2021 Annual General Meeting affirmed the goal of darker coat coloration. [55]

This outcome should be achievable by further selective breeding. A brown color between the dark stripes is absent in some plains zebras and others show this shadow striping to different degrees – suggesting that this coat coloration is under the control of multiple genes. A pronounced reddish-brown color is present in plains zebras in rare instances, as is clear in photographs from a variety of sources, including the Internet, and Cary described a plains zebra's hide with a, "copper color [that] is a strong, rich shade, not just a suggestion."[56] Genetic analysis of plains zebras with pronounced reddish-brown coloration could identify individuals that could be incorporated into the Quagga Project breeding stock, an approach that would be consistent with selective breeding procedures used throughout the Project.

Conventional breeding of animals and plants increasingly makes use of molecular biology. By collecting genomic information from many individuals within a species, it is possible to identify the several genes that determine a single characteristic such as milk yield in cows; breeders can use this information to decide which animals should be mated to yield offspring with the desired characteristic(s).[57] This approach of combining genomic information with conventional breeding has already been suggested to further reduce striping in Quagga Project zebras, and it would be feasible to pursue together the goals of reduced striping and a browner coat coloration.[58] Pedersen and his colleagues, who recently reported on the genomics of plains zebra populations, have proposed testing Quagga Project zebras, and the results could provide molecular information to guide further selective breeding.[59]

Registering and Selling Quagga Project Zebras

On May 6, 2008, the Quagga Project Association registered as a Section 21 non-profit company, which is equivalent to a 501(c)(3) organization in the United States and to some tax-exempt organizations in the United Kingdom.[60] In 2013, the Association registered the name "Quagga" as a certification mark; this move prevents unaffiliated individuals from breeding plains zebras and claiming that they are quaggas from the Quagga Project.[61] The "Rau Quagga" was approved as a recognized breed by South Africa's Ministry of Agriculture, Forestry and Fisheries. The Quagga Project has registered an independently regulated breed society, "The Rau Quagga Breed Society of South Africa" that includes all Rau quagga breeders and which will set breed standards.[62]

As a testimony to the success of the Quagga Project, three zebras with desirable coat characteristics fetched high prices at an auction in Stellenbosch, Western Cape on March 17, 2017 – a date that was almost thirty years to the day after the first zebras of the Quagga Project were captured. The occasion, which also featured sales of bonteboks, Cape mountain zebras, and kudus, attracted a sizeable number of people to an attractive rural venue.

"This is a very meaningful conservation project that brings you right here now," promised the auctioneer as he contemplated the auction of a Quagga Project stallion and two mares.[63] The auctioneer noted that they were in "uncharted waters" regarding the price the zebras would command. After lively bidding, the question was answered: each animal fetched R580,000 (approximately US$45,560), which was over fifty times the auction price of a fully striped plains zebra, although the price per animal was effectively less because both mares were pregnant.[64]

The outcome was a surprise to March Turnbull and Bernard Wooding of the Quagga Project, who were at the auction and were pleased when the bidding reached R580,000; they had initially thought this amount was the sum for all three animals and were delighted when they learned otherwise.[65] Their joy was not in the amount itself, but in what it could provide for the Quagga Project: funds to relocate zebras, and to buy needed items such as a camera with a telephoto lens and a GPS system to provide better photographic documentation of the animals together with records of their locations.

The auction provided an instance of different terms being applied to Quagga Project zebras: the auctioneer claimed, "They fall in the top 10% of quality in terms of the Quagga Project," although the Quagga Project described them as "Three (medium range) animals."[66] However, a screen beside the auctioneer described the animals as a "Quagga family group," and at times featured a single animal labeled "Quagga."[67] The Quagga Project website report of the auction was more circumspect in referring to the animals as "three unusual looking plains zebras."[68] The philosopher Delia Graff Fara argued that vagueness can be valid when the goals sought are imprecise, and the terms used – "quagga family group," "top 10% of quality," and "three (medium range) animals," – point to different understandings of the same animals by different people.[69] And the buyers probably believe that they have something more significant than "three unusual looking plains zebras."

In the Dr. Seuss poem "If I Ran the Zoo," the typical animals encountered in zoos are viewed as unremarkable and the plea is for novelty.[70] Partially striped brownish zebras with an intriguing backstory certainly provides that quality. The Quagga Project will sell more animals in the future, though March Turnbull thinks the March 17 auction prices were "an anomaly" and that future sales will not yield this sum.[71] Even if this prediction proves correct, it is clear that a market exists for these animals and the money raised in future sales could be important for the Quagga Project.

Quaggas Restored to the Veld

The auction affirmed the value of Quagga Project zebras and provided much-needed funds for reorganization in the form of a Core Herd located at four sites, each of which would receive an annual grazing fee from the Quagga Project. These locations reflect the diversity of land use in the Western Cape Province: Elandsberg (a farm with an adjacent nature reserve), the Nuwejaars Wetlands Special Management Area near Bredasdorp (a conservancy of several farms which have aggregated their marginal lands into an extensive nature reserve), Pampoenvlei (a private nature reserve of area 6.95 square miles or 1,800 hectares), and Vlakkenheuwel (a mixed use commercial farm that includes wheat fields). In June 2021, there were 106 Quagga Project zebras at these four locations.[72]

Since the March 2017 auction, the Quagga Project has moved many stallions between locations to prevent inbreeding and to put them with mares that ideally will bear Rau quaggas. During this reorganization of Core Herd animals, non-Core zebras – animals judged to be less desirable breeding stock – were sold or became the possessions of private land-owners on whose properties they were living. The herd that had lived on land at the edge of Table Mountain National Park was removed in November 2017 at the request of the SANPark authorities, who wanted to reduce grazing on the fynbos when this area was stressed by drought; the Quagga Project moved the best two stallions to Elandsberg and Pampoenvlei, enhancing the overall quality of their zebras at those locations.

The decision to move animals is based on factors such as their age and sex, and the carrying capacities of the locations where they live and where they are to be moved. As the Quagga Project is now restricted to four locations, avoiding inbreeding is more important than ever, and this objective is the province of the geneticist, Professor Eric Harley. I met with him during my visit to South Africa in April 2018 and saw the detailed records used to make decisions about relocation. Moving zebras requires permission from the conservation department and the state veterinarian, as these animals can carry and transmit African horse sickness.[73] Relocation is complicated and expensive but has been carried out successfully during the current reorganization.

March Turnbull and Bernard Wooding described their current procedure for moving zebras.[74] They carry out translocation at the coolest time of the year to avoid heat stress and time it so that animals are released at their new location with at least an hour of daylight remaining. The game capture team typically comprises twelve to fifteen people. Usually, the animals cannot be approached close enough on the ground to dart them and so a helicopter is required, which is a major expense. The veterinarian on board the helicopter carries photographs of the animal(s) to be darted, selects an animal and fires a dart containing a drug that sedates the animal for one to two hours. Bernard Wooding follows the animals in a 4×4 vehicle and watches the darted animal in the five to ten minutes when the drug is taking effect – a fraught time as the confused animal might stumble or hurt itself by running into a boulder or fence. Wooding radios the location of the unconscious animal to the ground team led by a ground manager.

During a period of what March Turnbull described as "well-managed haste," the members of the ground team confirm that they have the right animal, cover its eyes to protect them, examine it for worms and ticks, take a blood sample for DNA identification and to monitor for diseases, and insert a microchip. They may carry out other studies on the tranquilized zebra, for example, on one occasion a student studied blood oxygen levels during the period of tranquilization.

Five or six people carefully lift the tranquilized animal onto a stretcher and transfer it to a trailer – usually a small one to prevent the zebra from rearing up and hurting itself. If transporting more than one zebra, the animals have to be from the same breeding group or they might fight. The veterinarian gives the zebra an antidote to the sedative and when the animal recovers the journey can begin. At the destination, the zebra walks down a ramp to its new home.

In April 2018, I saw part of the Core Herd at Elandsberg, in the Riebeek Valley about ninety-five kilometers (fifty-nine miles) north-east of Cape Town. The Elandsberg Farm, which is next to the reserve, is managed by Mike Gregor, chairperson of the Quagga Project, and is a diversified venture of livestock (Cape buffalos, cattle, and merino sheep), wheat cultivation, and tourism. Visitors can enjoy lunch and a game drive that ideally includes sightings of zebras, and can stay at an upscale Victorian guest house, Bartholomeus Klip.[75] During my visit, the drought had almost drained the reservoir, but a well supplied ample water. March Turnbull, Bernard Wooding, and I drove in a safari vehicle on to the Elandsberg Reserve, a 4,000-hectare (15.44 square mile) area of natural veld extending to the Elandsberg mountains, which is managed in conjunction with CapeNature. Elandsberg and the neighboring sixteen farms collaborate in the management of the natural area at the base of the mountains. The vegetation of the Elandsberg Reserve is Swartland Alluvium Fynbos and Swartland Shale Renosterveld, a flora that is critically endangered; the most plentiful plants were restioids (a grasslike perennial) and bushes growing four-to-five feet high. The vegetation and terrain were probably the same as that experienced by quaggas when they lived here, and the animals we saw would also have been familiar to them: Cape mountain zebras, ostriches, bonteboks, and geometric tortoises, the latter a critically endangered species.

The Elandsberg Reserve had twenty-six Quagga Project zebras in three breeding groups and a bachelor group, but on that day they seemed elusive. Eventually, spotting a herd of nine zebras, we left the rough track and drove over stone-strewn ground, brushing against the fynbos bushes, which released an aromatic scent. As we neared the herd, Bernard and March identified the individual animals and Bernard took photographs of them. March observed that two of the animals, Nina and Lance, were together with all the enthusiasm of a parent who believes that his child has made a fine match.

As Quagga Project zebras are free ranging and select their own mates, it is important that all the animals should have the desired coat characteristics. By carefully observing the association of stallions with mares, and of mares with their foals in settled breeding groups, it is possible to identify the parents of each animal. DNA samples taken from tranquilized zebras have confirmed the accuracy of these visual observations. This system has worked well, but I enquired whether they had considered *in vitro* fertilization. Bernard replied that they tried to keep things simple and natural, but March was more open to the idea, remarking that "If someone told us that we could extract eggs from Nina [a Rau quagga with a beautiful brown coloration], do IVF [*in vitro* fertilization], and put embryos in horses to make 400 quaggas. We'd do that."[76]

Back at the small cottage that serves as the office for the Quagga Project, I admired the continued service of Alex, a member of the founder population, whose hide now provides a handsome rug, and enquired about Quagga Project zebras at other locations.[77] The owners of Melkplaas Farm, who acquired the three animals at the March 2017 auction, later bought additional Quagga Project animals and in April

2018 had twelve, which makes them the biggest breeder outside of the Quagga Project. As March Turnbull noted, "They will be absolutely on the inside in quagga breeding."[78]

Initially, SANParks and the Quagga Project collaborated, but SANParks has distanced itself from the Quagga Project. One issue was the unacceptable risk to the gene pool of the endangered Cape mountain zebra from mating with Quagga Project zebras to produce interspecific hybrids.[79] Commenting on this situation, Angela Gaylard of SANParks opined that Quagga Project zebras should live on farms, not with mountain zebras in national parks.[80]

March Turnbull observed that, "SANParks neither needs, nor wants, Quagga Project zebras."[81] There was regret in his voice, but he went on to observe that the "first prize" would still be to have Quagga Project zebras go into a national park: "If we had fifty brown pure-breeding quaggas, we would find someone to take them. We must find the right custodian." A population of fifty zebras would be large enough that inbreeding should not be a problem, but the challenge would be to find an area large enough to accommodate this many animals without exceeding the carrying capacity of the land. A SANParks reserve would be ideal, but losing the original collaboration with SANParks makes this goal difficult.[82] The Nuwejaars Wetlands Special Management Area is promising in this regard: some of the best Quagga Project zebras are already present here and the recent expansion of the land available to them could allow a population of fifty zebras, which would be the largest herd of these animals.[83] Nuwejaars borders Agulhas National Park and so it would be easy to move animals there if SANParks were to renew collaboration with the Quagga Project.[84]

Support for the Quagga Project

Rebreeding has involved many people and organizations acting with differing – though often overlapping – motives. Cape Nature Conservation (CapeNature), the South Africa Nature Foundation and South African National Parks (SANParks) provided crucial support, and many other individuals and organizations have contributed. Some donors would have responded to the 1989 fundraising appeal reproduced in Figure 8.1 in which quaggas are described, the scientific basis of rebreeding is explained, and a diagram shows the progression from plains zebras to a quagga-like animal. The human cause of extinction is noted, and the potential donor is urged to join with scientists and South African

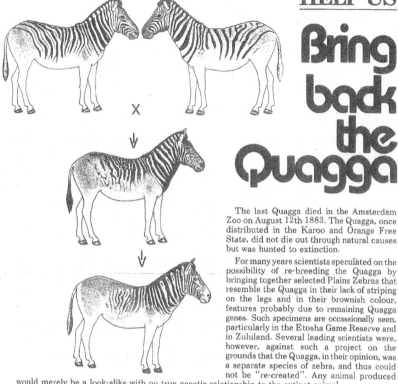

HELP US

Bring back the Quagga

The last Quagga died in the Amsterdam Zoo on August 12th 1883. The Quagga, once distributed in the Karoo and Orange Free State, did not die out through natural causes but was hunted to extinction.

For many years scientists speculated on the possibility of re-breeding the Quagga by bringing together selected Plains Zebras that resemble the Quagga in their lack of striping on the legs and in their brownish colour, features probably due to remaining Quagga genes. Such specimens are occassionally seen, particularly in the Etosha Game Reserve and in Zululand. Several leading scientists were, however, against such a project on the grounds that the Quagga, in their opinion, was a separate species of zebra, and thus could not be "re-created". Any animal produced would merely be a look-alike with no true genetic relationship to the extinct animal.

However, it has recently been shown by two groups of American scientists, working independently on Quagga tissue sent by the S.A. Museum and the Natural History Museum at Mainz, West Germany, that the Quagga was undoubtedly a subspecies of the Plains Zebra. This of course means that the main gene-pool is still available for re-breeding the Quagga without the introduction of any extra-specific genes. The DNA sequences of the Quagga, obtained from the dried muscle and blood, show virtual identity with sequences from living Plains Zebras.

As a result of this important finding, a committee of leading scientists from the S.A. Museum, the Universities of Cape Town and Stellenbosch and the Cape Provinicial Administration has been formed. The aim of this committee is to bring selected zebras to the Nature Conservation Station Vrolijkheid at Robertson and, through careful breeding, to retrieve the Quagga subspecies.

All four institutions provide scientific advice and assistance. The South African Nature Foundation has covered initial costs, and the Confederation of Hunters Associations of Southern Africa is contributing substantially. Much more money is needed for transport of zebras, fences, food, etc.

We appeal to you to contribute towards this project, which not only has great scientific interest but also ethnical value in bringing back an animal shamefully exterminated through short-sightedness.

Please send your contributions to: Quagga Experimental Breeding Programme, South African Museum, P.O. Box 61, Cape Town 8000.

Figure 8.1. "Help us Bring back the Quagga," a flyer produced by the Quagga Project in the late 1980s. Courtesy of the Quagga Project Association[85]

institutions in an ethical venture of scientific merit that will "bring back the quagga" and so will right a wrong. This appeal makes a persuasive case and undoubtedly was effective in fundraising.

Five different organizations supplied plains zebras to the Quagga Project, and others donated fodder and accommodations.[86] Even auditing and tax advice has been contributed.[87] Triggerfish Animation Studios, makers of the animated film, *Khumba*, which features a quagga-like zebra in the title role, gave a donation and a Rau quagga was named "Khumba."[88]

Many individuals, companies, and organizations provided financial support, including such varied donors as the Detroit Chapter of Safari Club International, the Friends of the South African Museum, and the American Long Ears Society. Contributions from these sources have been critical, as there has been no state funding, and expenditures are significant.[89] For example, the Quagga Project spent about R70,000 (approximately US$24,700) in 1993 to care for over twenty zebras.[90] Currently, expenses for the Quagga Project include grazing fees and the costs of relocating and caring for the animals. The Quagga Project has estimated that it will exhaust the funds gained in 2017 by the middle of 2021, but further sales could extend this date.[91] The Quagga Project has kept costs low by a mixture of fiscal responsibility, volunteer labor, and use of existing farms and reserves. Even if these locations do not charge commercial rates to house the zebras, some gain revenue from tourists who come to see the animals.

The Confederation of Hunters Associations of South Africa (CHASA) was moved by the moral appeal. When donating the first check from CHASA in 1987, Dormehl Vosloo commented, "Because the hunters were blamed for the fact that the quagga became extinct, we decided to support this venture."[92] CHASA was such an important supporter that Rau asserted that the Quagga Project would have failed in its absence.[93] Presumably, other supporters are similarly motivated to redress a wrong. This impetus might also be coupled with a wish to associate with a vigorous organization that has worthwhile goals – as is probably the case for the Wedderwill Wine Estate whose website proudly observes, "The owners of the Wedderwill Game Reserve are part of the quest of bringing back the Quagga."[94] In addition, some South African donors are likely motivated by national pride, as suggested by Pienaar, one of Rau's earliest supporters, who stated in 1977, "A project of this nature could have tremendous prestige value not only for the Parks Board but also for the country."[95]

The Quagga Project and De-extinction

Rau viewed breeding to produce quagga-like animals as a moral imperative and as a means of environmental restoration; both these motives
probably lie behind current de-extinction projects. The International
Union for the Conservation of Nature (IUCN) has issued guidelines
for such projects directed towards conservation and notes that the
resulting organisms will not be exact copies of the extinct species but
proxy organisms that would help to bring back lost biodiversity while at
the same time maintaining current biodiversity.[96] The techniques used in
de-extinction include biotechnology such as cloning and gene editing
but the initial focus in this account will be a comparison between
programs that use conventional breeding techniques to rebreed quaggas,
tortoises, and aurochs.

The selective breeding of cattle to produce animals resembling aurochs
that the Heck brothers began has taken on new life in several separate
projects that variously incorporate heritage breeds, Heck cattle, and
African Watusi cattle. The breeders select these animals for phenotypic
characteristics such as large size and long horns and their genetic similarity
with aurochs could be determined by comparing their DNA sequences
with the nuclear genome recovered from an animal that lived over 6,000
years ago.[97] The genomic outcome, arguably, is less important than the
creation of proxy organisms that resemble aurochs and that exert similar
effects on their environments.

The program that most closely resembles the Quagga Project is the
Giant Tortoise Restoration Initiative (GTRI) that is using selective
breeding to rebreed the Floreana tortoise (*Chelonoidis elephantopus*), a
Galapagos species that went extinct approximately 150 years ago.
Mariners who had captured these animals for provisions released some
on Isabela Island where some interbred with a closely related species.
Genetic analysis has identified twenty-three tortoises with Floreana
ancestry that the GTRI will use as a breeding stock; those offspring most
resembling Floreana tortoises will be returned to their native island.[98]
Given the longevity of these animals, those moved to Isabela Island may
be only one generation away from the original transplants and the
prospects of success seem high.

Rebreeding an animal by concentrating its genes within a population
has been used for red wolves (*Canis rufus*) in the United States. These
animals had been brought close to extinction by hunting and hybridization with coyotes. In the 1970s, more than 400 of these animals were
captured in the southeastern United States and their morphology

evaluated. Forty-three of these animals were judged to be red wolves and were bred in captivity. Careful examination of these animals and their pups detected some red wolf–coyote hybrids, and so there was further selection that resulted in fourteen individuals classified as "pure red wolves" which were the founder population for a captive breeding program that released red wolves into a habitat having prey but few coyotes.[99] Although their new habitat differed from their original one, the population grew and by 2002 there were approximately one hundred red wolves. Major challenges to this successful rebreeding project have been loss of animals to vehicles and hunting, and the opposition by some people to having a wild carnivore restored to the land.[100] The red wolves never ceased to exist and so the project did not involve de-extinction, but it is included here as a parallel example to the Quagga Project of concentrating particular genes within a population to create a desired phenotype based on morphological characteristics.[101] Other similarities include release into an environment that differs from the original (because of human activities and climate change in the case of Quagga Project zebras) and possible problems caused by interbreeding – with coyotes in the case of red wolves and other plains zebras in the case of Quagga Project zebras.

All the previous examples made use of conventional breeding techniques, but what has been termed "conservation cloning" could be used when living cells of extinct organisms are available. This technique is based on the work of John Gurdon who in the late 1950s transferred the nucleus of a fully differentiated tadpole cell to an enucleated egg cell of the same species – a technique termed somatic cell nuclear transfer.[102] The transplanted nucleus dedifferentiated, and its cell divided to form an embryo that developed into a frog. This experiment showed that the nuclei of differentiated cells are totipotent, that is, in forming a specialized cell (in this case, one from the intestinal epithelium), they maintain all of their genes and so retain the capacity to make every cell type of an organism. Totipotency was famously demonstrated in mammals with the birth of Dolly in 1996. A differentiated cell removed from a sheep mammary gland was fused using pulses of electricity with an egg whose nucleus had been removed.[103] Cell divisions of the hybrid cell formed an early embryo that was implanted into a sheep uterus where embryonic development occurred normally and a clone of the mother sheep was born.

This experimental work paved the way for cloning of horses, cows, pigs, deer, monkeys, and dogs. These successful ventures suggested the use of similar techniques for endangered and extinct species, and a Przewalski's horse was recently born from cells cryopreserved in 1980.[104] Conservation cloning is possible in those cases where far-sighted people have removed cells from endangered animals and established tissue cultures of them. One such collection of animal germ and somatic cells is the "Frozen Zoo" at the San Diego Zoo Institute for Conservation Research.[105] Extinct species whose somatic cells live on in tissue culture include the gastric brooding frog (*Rheobatrachus silus*), and a wild goat, the Pyrenean ibex (*Capra pyrenaica pyrenaica*). Theoretically, a cell of an extinct animal could be fused with an enucleated egg cell of a related species, and if an early embryo were formed it could be transplanted into the uterus of a surrogate mother of a closely related species where the embryo could complete its development.

Cloning makes use of complex procedures. Dolly was the sole animal to be born from many attempted nuclear transfers, and this outcome was in a species whose reproductive biology was well known. Conservation cloning faces additional technical challenges: the biology of the organism is probably less well known, and its nucleus is transplanted into the enucleated egg of a different species that may not be compatible. Should an early embryo develop, it will be transplanted into a surrogate of a closely related species that may reject it.

In the sole instance where a cloned extinct animal was born, a Pyrenean ibex died after several minutes of life and was found to have malformed lungs. Even if healthy animals are born and de-extinction becomes a reality, there would be additional challenges: how would the animal learn the behavior of its extinct ancestors, and would it have access to other genetically different animals of the same species with which to breed? The latter consideration could only be met if tissue cultures had been established from multiple animals of the species before extinction. Finally, de-extinct animals would have to exist in an environment different from that of their ancestors: a world with habitats altered, more human activity, and climate change. Consequently, while it is possible that scientists will clone extinct animals, they face formidable challenges.

A final possibility is gene editing used for de-extinction of animals where viable nuclei are not available but whose genomes have been sequenced from recovered preserved tissues.[106] Identification of genes that gave the extinct animal its morphological, physiological, or behavioral characteristics would be followed by synthesis of those genes *in vitro* and insertion into precise sites in the nuclear genome of a closely related animal. The genetically altered nucleus would then be inserted into the enucleated egg cell of a closely related animal. The subsequent stages required to yield a transgenic proxy would be the same as described for conservation cloning and the same challenges would be present, but these would be in addition to difficulties involved in identifying the appropriate genes and gene editing at several – possibly many – sites in the genome. Nonetheless, the organization Revive and Restore is working on four de-extinction projects using gene editing, including passenger pigeons and woolly mammoths. Ben Novak, the Lead Scientist at Revive and Restore has estimated that gene editing of a plains zebra could result in birth of an animal with quagga-like coat coloration in less than five years and at a lesser cost than incurred by the Quagga Project.[107] It will be fascinating to see if anyone accepts this challenge.

For the foreseeable future, the only zebras with quagga-like coat coloration will be those produced by selective breeding. Contrasted with the highly technical approach of clonal conservation, the Quagga Project recalls an earlier age when people improved animal breeds by careful selection of animals to have the qualities they valued.

Public Meanings of the Quagga Project

With the transition to democracy in 1994, the Truth and Reconciliation Commission (TRC) provided opportunities to think about past wrongs and to try to redress them. Perhaps this zeitgeist inspired Rau when he framed quagga rebreeding as "our moral duty to rectify that mistake."[108] Acknowledging the wrong is only the first step. Rectifying it requires compensatory action, and that is where bills mount up. To support the effort, the Quagga Project has drawn on many people's sense of duty to repair what humans have destroyed, and this aspect, in itself, is remarkable.

The New South Africa is sometimes called the "Rainbow Nation," but this diversity is less apparent in the Quagga Project where most names conferred on animals are drawn from a sliver of South African society: Khumba is an exception amid Alex, Margaret, Rebecca, Henry, etc. The video of the March 17, 2017 auction of Quagga Project zebras at Stellenbosch shows a demographic that is mainly white and relatively wealthy.[109] The same is probably true for those visiting wildlife ranches where a two hour-long ride to view Rau quaggas costs R2500 ($170) for a party of three – a considerable sum for many South Africans. As in many other countries, convincing citizens about the importance of conserving animals will be challenging when most people cannot afford to visit them in their habitats and so find it difficult to relate to the goals of their conservation.

The Quagga Project has been the subject of television documentaries and articles in journals and newspapers. Media attention has been favorable, with headlines such as "Karoo welcomes quagga back from the dead" and "Brave quest of Africa hunt: bringing back extinct quagga."[110] The optimistic tenor of some reports might lead some readers to believe that extinction is reversible; however, the Quagga Project specifically addresses this issue by pointing out that losing a species is "irreversible" and notes that quaggas were not a separate species, but a subspecies of plains zebras.[111] This point was also emphasized in the 2008 Voice of America News documentary that, however, bore the inappropriate title, "Rebuilding a Species".[112]

These reports tell an uplifting story that is a welcome change from descriptions of climate change, species extinction, and habitat degradation. Yet, all of these harmful processes are occurring. The danger remains that the rebreeding of Rau quaggas might lead people to become less concerned about extinction. Rick de Vos voices a similar concern: mourning the world that was lost with the extinction of quaggas, he finds it ironic that the presence of Quagga Project zebras in the Karoo overshadows the loss of the organisms that helped to shape the Karoo environment.[113] His regrets echo Macnab's poem, "The Quagga's World," in which microorganisms, fungi, flies, and worms are appreciated as part of this environment – affecting quaggas and, in turn, being influenced by them.[114]

What Has Been Gained?

It was overly ambitious to envision having "500 quagga-like zebras by 2020" and unrealistic to assume it would be possible to breed animals with reddish muzzles.[115] Despite not meeting all the goals laid out in its original *Management/Action Plan*, the Quagga Project has been successful in at least three respects. First, animals similar in coat coloration to quaggas now exist in South Africa thanks to visionaries, philanthropists, and careful breeding; this return is worthy of celebration. Second, the Quagga Project affords a well-documented instance of the power of selective breeding: the reduced striping over five generations provides quantitative data as a contemporary example of the type of selection described in Darwin's *The Variation of Animals and Plants under Domestication*.[116] Finally, the Quagga Project provides a model for how people with different perspectives can work together to bring about beneficial change; this aspect is the broader lesson of the Quagga Project.

9 · *Identity and Conservation*

The Project's single objective, unwavering over three decades and finally made possible by the first successful sale of Quagga Project animals, is to establish and protect in perpetuity, a core herd of the very best animals in the project. The herd will be continually "improved" until they are indistinguishable from the original quagga when they will hopefully "rewild" game reserves across their original range.

Anon, 2017[1]

Historically, people debated the nature of quaggas. Were they a separate species, or just a less-striped, browner form of plains zebras? Did they resemble horses or donkeys? Were they a nuisance animal that deprived farm animals of water and forage, or a protector of livestock and a beast of burden? Were they a source of meat and hides or a wild animal that should be allowed "to range unmolested"?[2] Identity is an issue, too, for the rebred zebras of the Quagga Project: to what extent do they resemble quaggas and do similarities in coat coloration justify the assertion that, "We see the project as comparable to the conservation of endangered species ..."?[3] And even if the claim that rebreeding equates with conservation is rejected, can the introduction of Quagga Project zebras into habitats once occupied by quaggas aid the conservation of these environments?

Are the Rebred Animals Quaggas?

The Quagga Project hopes that the rebred zebras will be "indistinguishable" from quaggas. Yet, just because a man looks like Albert Einstein does not mean that he possesses Einstein's creativity.[4] In this instance, selection for coat coloration alone would not retrieve other characteristics of quaggas that were not present in the plains zebras from which they were rebred.[5] Rebreeding extinct animals requires a thorough knowledge of their characteristics and yet much of this information is lacking for quaggas: there never will be opportunity, for example, to observe in detail their interactions within a herd or to record their vocalizations and to compare these with other subspecies of plains zebras. Similarly, we know nothing about their physiology, and so questions about adaptations to the cold and aridity of the Karoo remain unanswered.

The Quagga Project addresses whether quaggas had unique physiological adaptations unrecoverable by selective breeding by asserting, "since extant Plains Zebras occupy habitats of similar degree of aridity to those of the Quagga, there is no sound reason for proposing significant adaptive features of the Quagga to its original habitat."[6] However, a single "original habitat" never existed: quaggas lived in environments varying in topography, vegetation, and climate extending from renosterveld near Cape Town to the grasslands of the Orange Free State. Moreover, during the hundreds of thousands of years that quaggas evolved their distinctive coat coloration, their territories varied in temperature and aridity. Genetic differences between quaggas and other plains zebras resulting in reduced striping and brown coat coloration might be accompanied by genetic variations for other morphological characteristics, and for behavioral and physiological features. The Quagga Project notes this lack of information about quaggas:

It has been argued that there might have been other non-morphological, genetically-coded features (such as habitat adaptations) unique to the Quagga and that therefore, any animal produced by a selective breeding programme would not be a genuine Quagga. Since there is no direct evidence for such characters and since it would be impossible now to demonstrate such characters were they to exist, this argument has limited value. The definition of the Quagga can only rest on its well-described morphological characteristics and, if an animal is obtained that possesses these characters, then it is fair to claim that it is a representation of, at least, the visible Quagga phenotype.[7]

The Quagga Project's assertion that, "The definition of the Quagga can only rest on its well-described morphological characteristics" overlooks the fact that these same characteristics led Gray in 1825 specifically to distinguish plains zebras as a separate species from quaggas which had been described and named earlier.[8] For over 150 years, *Equus quagga* and *Equus burchellii* remained as separate species. It was analysis of proteins and DNA that defined quaggas as a subspecies of the plains zebra, although even when molecular knowledge was available some biologists continued to distinguish quaggas from plains zebras based on morphology.[9]

Quaggas' distinctive morphological characteristics go beyond coat coloration; they also include cranial structure and dentition that, as discussed in Chapter 7, vary between quaggas and plains zebras.[10] In the absence of selection for these features, Rau quaggas will not resemble extinct quaggas, but will have the cranial structure and dentition of plains zebras. "It would be comforting to think that its essence lives on in the plains zebra," was Klein and Cruz-Uribe's conclusion about quaggas after comparing their cranial measurements with those of mountain zebras and plains zebras "but our craniometric study suggests that something more unique may have been lost."[11]

Sexual dimorphism is another important morphological characteristic to consider: among quagga taxidermy specimens, mares were significantly longer and slightly taller than stallions.[12] As there has been no selection for this sexual dimorphism in body size, Quagga Project zebras will not resemble quaggas but will probably be similar to other plains zebras in which mares are slightly smaller than stallions.

Pedersen et al. used the quagga genome sequenced by Jónsson et al. as part of a study to show the divergence of plains zebras probably from the Zambesi-Okavango wetlands about 367,000 years ago, culminating in quaggas and Burchell's subspecies in southern Africa and a genetically distinct population in present-day Uganda. This important paper concluded: "We show evidence for inclusion of the extinct and phenotypically divergent quagga (*Equus quagga quagga*) in the plains zebra variation and reveal that it was less divergent from the other subspecies than [was] the northernmost (Ugandan) extant population."[13] The Quagga Project's Coordinator's Report commented on this paper, "This report suggests that there really doesn't seem to have been anything else [other than coat coloration] different about the quagga (genetically) and does not support the argument that the Quagga Project may be missing some mysterious and lost DNA that defined the quagga."[14] However, this assertion goes beyond the conclusion of the paper that acknowledged a divergence between quaggas and other plains zebras while also noting that the Ugandan population was an outlier.

In fact, the study by Jónsson et al. of a quagga's genome showed that there *were* significant genomic differences between quaggas and *Equus quagga boehmi*. In particular, there had been positive selection for the *SEMA5A* gene in quaggas and positive selection for a different gene, *SLC9A4*, in *Equus quagga boehmi*.[15] To appreciate the importance of this discovery, it is helpful to consider two well-studied examples of positive selection for genes in humans: lactose tolerance and resistance to malaria.

In the former, a mutation keeps the lactase gene active after weaning and so enables adults to digest milk.[16] This benefit presumably aided survival when food supplies were short and so had a selective advantage for people with the mutant gene. Another positive selection was for the sickle-cell allele that produces red blood cells resistant to the *Plasmodium* parasite – enhancing survival in areas where malaria is prevalent.[17] The significance of positive selection for different genes in *Equus quagga boehmi* and in quaggas may become clearer in the future, but for the moment these genomic differences refute the statement that quaggas differed from other zebras only in coat coloration.

The Quagga Project has acknowledged these genomic differences between quaggas and other plains zebras and the implications of this information for the characteristics of rebred animals: "One interesting point is relevant to the question as to whether the quagga had any special attributes (other than its pelage) for which we are not selecting, is an observation that there appears to have been some positive selection in the quagga for a particular gene which might be involved in behaviour modification."[18] This acknowledgment notwithstanding, at times the Quagga Project seems to regard some of their animals as *being* quaggas. Under the heading "View Quaggas," their website offers this direction, "To view Quaggas that have been reintroduced into reserves, contact Elandsberg Nature Reserve or Bontebok Ridge Reserve near Wellington, or Nuwejaars Wetland near Bredasdorp."[19] A similarly expansive claim occurs elsewhere on the website: "since the coat-pattern characteristics are the only criteria by which the Quagga is identified, re-bred animals that demonstrate these coat-pattern characteristics could justifiably be called Quaggas."[20] There are even instances of the trinomial name being conferred on Quagga Project zebras as in this response by the writer of the Coordinator's Report to whether or not quaggas are extinct, "Quaggas are indeed not extinct, *Equus quagga quagga* is its Latin name."[21]

March Turnbull of the Quagga Project is more careful in his characterization: "We are not arrogant. We don't know that these zebra are quaggas. We only know what the extinct quagga looked like, not what sounds they made or what behavioural characteristics they exhibited."[22] The Quagga Project website has a similarly cautious assessment which differs from some of its other statements:

It has been decided that when an animal is achieved which has no scorable stripes on the hind part of the body, and no stripes on the legs, then it qualifies.

Accepting that we are only selecting for this set of attributes of the original quagga, and not for any other genetic features which may have been possessed by the original quagga population, we will then term this a "Rau quagga ..."[23]

This statement is reasonable: it makes coat coloration the sole criterion of animals which are distinguished by the qualifier "Rau," and it is open to the possibility that they are genetically different from quaggas.

South African National Parks (SANParks) shows similar restraint on their websites in describing Quagga Project animals in the Karoo National Park as, "Burchell's Zebra with Quagga bred characteristics," and in the Addo Elephant National Park as "The Burchell's zebra, many with the pale rumps reminiscent of the extinct quagga."[24] With regard to identity, my approach has been to refer to all rebred plains zebras as Quagga Project zebras, unless they have been designated as Rau quaggas, in which case I use this term.

Rewilding with Quagga Project Zebras

Rewilding is the (re-)introduction of one or more organisms to improve the operation of an ecosystem.[25] Once ecosystems have been rewilded, the intention is that the introduced organisms – often apex predators or ecosystem engineers – will exist with the minimum of human intervention. Rewilding with rebred animals is challenging as the original biota will have changed since extinction, with some organisms having increased or decreased in numbers or become extinct.[26] The physical environments where extinct organisms once lived will have become different and will continue to change because of human action.

The reintroduction of wolves (*Canis lupus*) into Yellowstone National Park, USA, in the late twentieth century is a well-known example of rewilding. The extirpation of these apex predators from this area led to increased numbers of elk that browsed on vegetation, including young woody plants. The effects of wolf reintroduction are complicated by other factors that occurred at about the same time, including climate change, drought, and increases in the populations of bears and beavers.[27] Nonetheless, wolves preyed on elks and with their browsing reduced there have been increases in woody plants, including aspen, cottonwood, and willow.[28]

The reintroduction of beavers to their former habitats provides European examples of rewilding. Beavers are ecosystem engineers: they fell trees and cause changes in hydrology by damming streams and digging channels. Impounded water will provide environments for some aquatic organisms but will eliminate others such as trees killed by water logging of their roots; this outcome is not necessarily negative because dead trees provide habitats for many organisms. Hunting of beavers for pelts or because people perceived them as a nuisance changed the original landscape. People reintroduced Eurasian beavers (*Castor fiber*) into Scotland in the early twenty-first century, over 400 years after their extirpation from that country. They expected that wetland restoration would kill trees but would provide habitats for some aquatic organisms. Some farmers opposed this move and killed the animals but reintroduction created riparian habitats and increased biodiversity; now beavers have legal protection in Scotland and probably their range will increase.[29] In 1957, American beavers (*Castor canadensis*) introduced into a boreal Finnish landscape devoid of native beavers since 1868 helped to restore riparian environments.[30] Although the animals and the environments are different, introduction of beavers, wolves, and Quagga Project zebras should provide some degree of environmental restoration.

Much rewilding in South Africa has occurred on new reserves, often converted from farmland.[31] The organisms that are re-introduced can be proxies of the original inhabitants and so people could stock a reserve with any subspecies of plains zebras and presumably their grazing would have a similar effect to that of the zebras that once lived there. Rau, however, criticized instances where people had stocked South African reserves with plains zebras imported from elsewhere in Africa that differed in coat coloration from the native Burchell's subspecies; he advocated removing "former indiscriminate reintroductions" and supplanting them with "individuals which are closer to the original population."[32] Following this principle, Quagga Project zebras are appropriate for introduction into the Karoo.

The release of Quagga Project zebras to locations where quaggas once lived is fitting both ecologically and symbolically. However, animals have cultures which are learned from conspecifics and which evolve over time.[33] That culture ended with extinction and is unlikely to be present in the same form in Quagga Project zebras. Perhaps, too, there was a shared way of life developed over many generations with the ostriches, wildebeests, and other animals with which quaggas grazed. If so, it is unlikely to be precisely restored, even if the same assemblage of animals that once interacted now live together again.

Another issue is whether there will be enough Quagga Project zebras to help rewild ecosystems whose native plants were affected by the extinction of quaggas and the reduction in numbers of other native herbivores?[34] Quaggas played an important ecological role and their extirpation certainly affected their former habitats.[35] Nonetheless, the assertion that, "The last, best hope for the fragile Karoo may be that one day it will be known as the place where the quagga was exterminated, only to rise again" is overly optimistic.[36] That region now differs significantly from what it was in the nineteenth century and earlier, and so the benefits of grazing by Quagga Project zebras cannot offset the effects of climate change, mining, fracking, and fences that disrupt migrations, etc.[37] Even so, it should be possible to find an area within the Karoo where these animals could live and have their effects on that local environment. Grazing by fully striped plains zebras would have a similar outcome, of course, but would not have the same symbolic significance. As Bernard Wooding commented, "If we didn't have quaggas at Elandsberg [Reserve], we would have normal [plains] zebras, but quaggas give a better story to tell to visitors." March Turnbull agreed, "It's the same animal on the same veld."[38]

Quagga Project and Conservation in South Africa

Professor Michael Cluver viewed the Quagga Project as a form of genetic conservation since it retrieves some genes of quaggas.[39] Reinhold Rau also asserted that the Quagga Project is "comparable to the conservation of endangered species and/or the re-introduction of rare species into areas where they have become extinct."[40] However, the 1975 judgment of the Natal Parks Board that attempts to rebreed quaggas had "very dubious conservation value" and the critique by the authorities in Namibia a decade later that the project was "of little importance to conservation" seem to argue that recovery of a coat color phenotype does not rank high among the many other pressing needs of South African wildlife.[41] Opinions vary: Don Boroughs wrote that Rau's interest in stocking the Karoo National Park with Quagga Project zebras, "makes some conservationists cheer and others cringe."[42]

The question of whether selection for coat coloration by the Quagga Project is conservation or not might be complicated by recent selective breeding in South Africa for wild animals with colors that would only rarely occur in nature. Conservationists with their concern for authenticity, rather than novelty, are opposed to breeding color variants of wild animals and possibly some people might place rebreeding for quagga coat coloration into this category. The Endangered Wildlife Trust defines the production of color variants as "the selective breeding of individuals with unique pelage colours or patterns, so as to promote that pelage."[43] Although this definition seems applicable to selection of plains zebras for quagga-like coat coloration, Andrew Taylor of the Endangered Wildlife Trust has indicated that the Trust's statement had been in response to the selective breeding by wildlife ranchers of color variants such as black impalas and golden wildebeests, not Quagga Project zebras.[44]

The 2016 Red List of Mammals of South Africa, Lesotho, and Swaziland expressed concern that some breeders were selecting for particular color morphs of plains zebras.[45] Keenan Stears, an author of the report, assured me that the concern was the deliberate production of color variants – including animals with quagga-like reduction in striping – by inbreeding plains zebras on wildlife ranches. He observed, "In reality, the threat to wild populations to create colour morphs likely comes from unregulated breeding for hunting rather than the Quagga Project."[46]

There are many conservation challenges in South Africa – a country of high biodiversity with many endemic species, some of which are threatened with extinction. In 2016, a survey of mammals of South Africa, Lesotho, and Swaziland found that 17 percent of taxa were threatened with extinction and 10 percent were near threatened.[47] More recently, biologists have estimated that extinction endangers 16 percent of South Africa's plant species – a percentage that has increased since 1990.[48] Native plants in South Africa have an array of challenges including climate change, fires, overgrazing, illegal collecting, introduced species, and the expansion of agriculture, industry, and human populations. Few of these plants would attract significant conservation funding and the same is true for endangered fungi and for many taxa of invertebrates. Intricate and imprecise factors, including biopolitics, are often in play when choosing conservation goals.[49]

Money for conservation is always tight and a necessary question for any wildlife project is whether the funds could be used to greater effect elsewhere. This issue was central to considering de-extinction of Australian and New Zealand animals: aside from the considerable expense of recreating a species, the costs of merely maintaining that species would use funds that could otherwise support actual conservation projects and so de-extinction might lead to an overall reduction of biodiversity.[50] For the Quagga Project, however, the maintenance costs are modest: suitable grazing habitats and the absence of other plains zebras to avoid dilution of quagga-like coat coloration in any offspring. Owners of these zebras can offset these expenses in those situations where tourists pay to see the animals.

The question of whether funds used by the Quagga Project could have been invested in preserving endangered species and their habitats has been raised on several occasions.[51] Of course, donors are free to support both the Quagga Project *and* conservation efforts. Yet a project involving zebras, attractive creatures with esthetic and corporeal charisma, will garner support that might not be forthcoming for many other organisms viewed as being less captivating.[52] Additionally, the Quagga Project is presumably appealing to many donors because it incorporates the idealistic theme of bringing back the phenotype of an animal thought to be extinct.

There are many endangered organisms that could have benefitted from some of the funds used by the Quagga Project but the one that stands out is the Selous' zebra, a "rare, morphologically-unique population" of the plains zebra that some biologists treat as a separate subspecies, *Equus quagga selousi*.[53] In neighboring Mozambique, less than fifty of these animals remained in 2000, reduced from over 20,000 in 1970, and a recent survey showed their numbers at less than 2 percent of historical populations.[54] Many of these animals were casualties of the Mozambican civil war when they provided food for hungry people and soldiers, but hostilities ended in 1992 and safari operators continued to hunt the zebras until 1999.[55] If people had established a captive breeding program for a few of these animals supported by some of the money donated to the Quagga Project, it is possible that this zebra would not now be so rare.

March Turnbull of the Quagga Project has given a lot of thought to the question of opportunity costs and is emphatic in his claim that it is unlikely "any of the money put into this project is available for conservation in other ways."[56] Additionally, land inhabited by Quagga Project zebras has benefited the conservation of other organisms, for example, both the Elandsberg Reserve and the Bontebok Ridge Reserve protect renosterveld and fynbos vegetation, together with the endangered animals that live on this vegetation, such as geometric tortoises.[57] The Quagga Project has introduced their zebras into the Nuwejaars Wetlands Special Management area which includes lowland fynbos and renosterveld habitats that are both rich in biodiversity and endangered species. The Nuwejaars Wetlands Special Management area supports other reintroduced animal species that once existed in this location – including buffalos, hartebeest, and hippopotami. These rewilding programs not only return animals to their original habitats but can also help restore vegetation damaged by removal of indigenous herbivores.

Finally, it bears noting that people active in the Quagga Project are often engaged in conservation activities. Reinhold Rau helped preserve geometric tortoises and Cape clawed frogs.[58] Eric Harley uses molecular biology to study the taxonomy of the white rhinoceros (*Ceratotherium simum*), which provides important information for their conservation.[59] March Turnbull was founding director of the MAPA Project, which has mapped African conservation sites. Such involvements show an active concern for maintaining existing biodiversity.

The Quagga Project as a Model

The Quagga Project provides a model for how a heterogeneous assortment of people and organizations with different agendas can work together successfully for a common cause. The rebreeding effort depended on several distinct components beginning with taxidermy specimens that generations of museum personnel had cared for and that ultimately provided Rau with tissues. He, in turn, recruited molecular biologists whose initial interests lay not in quaggas *per se* but in whether they could extract and sequence DNA from an extinct animal. These successful genomic studies provided key evidence for taxonomists whose reclassification of plains zebras then established the biological justification for rebreeding. Different specialists, with their varied interests, were all essential to showing that selective breeding was feasible. Seen in this way, the Quagga Project is a "boundary object," a term originating in a study of Berkeley's Museum of Vertebrate Zoology that analyzed how the museum staff, researchers, and the visiting public worked with each other even though they held different goals.[60]

The process of rebreeding was itself a boundary object as new benefactors – both people and organizations – became involved, including the dedicated people who have worked on the Quagga Project, cared for the animals, and have provided support in the form of donations or land where the zebras can live. As a whole, this enterprise presents a model that conservationists could apply effectively to the preservation of endangered species and their habitats.

Grevy's Zebra Trust

Conservation of Grévy's zebras by the Grevy's Zebra Trust based in northern Kenya provides another example where a range of people and organizations function together in imaginative ways.[61] Grévy's zebras are the largest and most striped of zebras; unfortunately, they are also the most endangered because of illegal hunting, loss of water and grazing, and climate change. Once widespread across northern Kenya and southern Ethiopia, their range has been reduced and some animals exist in isolated populations – a situation that can lead to inbreeding and loss of genetic diversity that adds to the factors imperiling their survival. The Grevy's Zebra Trust, like the Quagga Project, depends on people who care for animals and their environments. Its mission statement, "The Grevy's Zebra Trust conserves the endangered Grevy's zebra and its fragile habitat in partnership with communities" summarizes an approach that involves almost 100 indigenous people in collecting information about the animals and in designing strategies to conserve them and their environments. Use of visual data sheets allows non-literate people to participate in data collection and, aided by literate partners, some of these people have learned to write. Improved literacy and income from salaries are welcome additional benefits.

Local communities involved with the Grevy's Zebra Trust provide key information from reporting on water sources to aiding individual animals, for example, in rescuing zebras trapped on mud flats after seasonal flooding. Citizen science provides data on the land, people, livestock, and wildlife that communities deliberate at workshops before deciding on conservation strategies. Grevy's Zebra Trust involves stakeholders on every level from that of local communities to national leaders and has established collaboration with international associates, including Princeton University, the Saint Louis Zoo, the San Diego Zoo, and the Wildlife Conservation Network. Since its founding in 2007, the Grevy's Zebra Trust has already accomplished much and, like the Quagga Project, has the supporters, donors, and partner organizations to ensure its success.

Gains, Losses, and Hope

The last fifty years have seen both gains and losses for zebras. Cape mountain zebras have increased in numbers and Rau quaggas have come into existence. At the same time, populations of Selous' zebras in Mozambique and Grévy's zebras in Kenya and Ethiopia have suffered alarming decreases. The causes of decline affect many other species: habitat destruction, climate change, and loss of organisms because they were valued for the wrong reasons or not valued enough.

In 1937, the biologist Otto Antonius concluded his survey of the geographical distribution of equines: "Generally speaking the extinction and restriction of the Equidae is no more than one act of the great ragedy happening in our days: the end of the age of mammals."[62] Read at the time, extending a conclusion about equines to an entire class of animals might have seemed an unjustified generalization. Seen from our current perspective in the sixth mass extinction, restricting "the great tragedy happening in our days" to mammals is too limited in an age where many varied organisms are becoming rare or lost altogether.

People need to be aware of this tragedy but must not give way to despair. With apologies to Emily Dickinson, hope is not only "the thing with feathers" but is also a beautiful equine. How inspiring that these creatures exist in a world of change and loss. Quaggas were not widely known in their lifetimes and have been extinct for a century and a half, yet their rebreeding evokes the same optimism as other restoration attempts. "People need hope. People need symbols," said an ecologist contemplating the restoration of American chestnut trees that may once again provide splendor, wood and nuts to the forest landscape.[63] Rau quaggas are similar symbols of hope.

Appendix 1
Early Illustrations of Quaggas[1]

Appendix 1 lists illustrations of quaggas made prior to their extinction. Images that were copied from earlier illustrations are not included, nor is artwork dated after 1883 as these images were based on previous work or taxidermy specimens.

n.d.: Rock engraving from Wildebeest Kuil near Kimberley, Northern Cape Province; see Figure 2.2.

n.d.: Rock engravings from Kinderdam near Vryburg, North West Province.

n.d.: Rock engravings near Britstown, Northern Cape Province.

1751: George Edwards, see Figure 2.4.

1777: Robert Gordon, see Figure 2.3.

1780: Aert Schouman, see Figure 2.6.

1793: Nicolas Maréchal, see Figure 2.7.

1804: Samuel Daniell, see Figure 2.5.

1817: Jacques-Laurent Agasse.

1820: Christian Wilhelm Karl Kehrer.

1821: Jacques-Laurent Agasse, see Figure 2.9.

1824: James De Carle Sowerby.

1836: W. Cornwallis Harris, see Figure 2.10.

1841: C. Hamilton Smith.

1844: Johann Andreas Wagner.

1850: Waterhouse Hawkins, see Figure 2.8.

1858: Harrison Weir, see Figure 0.1.

1860: Theodor Zimmermann.

1863: Frank Haes.

1864: Frank Haes, see Figure 5.1.

1864: Gustav Fritsch, see Figure 2.1.

1870: Frederick York.

Appendix 2
Records of Quaggas Kept in Europe

Location	Dates	Details
NETHERLANDS		
House of Orange	Late 1740s	Several quaggas were kept by King Willem IV.
House of Orange	1780	Quagga captured by Robert Gordon.
Amsterdam Zoo	1851–1853	Mare purchased from Lord Derby's menagerie at Knowsley. She bore a hybrid foal from mating with an onanger.
Amsterdam Zoo	1872–?	Sex and date of death are unknown.
Amsterdam Zoo	1867–1883	Mare that was the endling of the species. Her hide is the Amsterdam quagga.
BRITAIN		
Prince of Wales' menagerie at Kew Palace	1751	Mare painted by George Edwards in 1751 (Figure 2.4) which is the type specimen of *Equus quagga*.
Dalmahoy House near Edinburgh, Scotland	1814–?	Stallion owned by Lord Morton that produced a foal with a horse. Painted by Agasse in 1821 (Figure 2.9).
Tower of London Royal Menagerie	1817–1820	Two stallions. They pulled the carriage of Sheriff Parkins in London. Agasse painted one in 1817.
The Exeter Change, London	1824	Mare.
Royal Menagerie in Windsor Great Park	1826–1830?	Stallion, mare, and quagga of unknown sex.
Owston, Yorkshire	1830	A mare that produced two foals by a donkey.
London Zoo	1831–1834	Mare.
London Zoo	1851–1872	Mare photographed by Haes (Figure 5.1) and York. Skeleton is in the Peabody Museum at Yale University. Hide is the Edinburgh quagga.
London Zoo	1858–1864	Stallion donated by Sir George Grey, Governor of Cape Colony (Figure 0.1). Skeleton is in the British Museum of Natural History. Hide is probably the Wiesbaden quagga.

(*cont.*)

Location	Dates	Details
Lord Derby's menagerie at Knowsley, Lancashire	1847–1851	Two painted in 1850 by Waterhouse Hawkins (Figure 2.8). Sex of one is unknown, the other was a mare that was sold to the Amsterdam Zoo in 1851.
FRANCE		
Palace of Versailles	1784–1794	Stallion belonging to King Louis XVI and kept at Versailles. Painted by Maréchal in 1793 (Figure 2.7).
Paris Botanical Gardens	1794–1798	Stallion moved from Versailles. Its hide is now the Paris quagga.
AUSTRIA		
Schönbrunn Gardens, Vienna	1786–1798	This stallion might have been the one caught by Robert Gordon that was formerly at the menagerie of the Prince of Orange.
GERMANY		
Royal Menagerie at Karlsruhe	1817–1818	Two quaggas.
Bavaria	1818–?	Two quaggas which the King of Bavaria purchased from Karlsruhe.
Berlin Zoo	1863–1867	Mare. Her hide is the Berlin quagga.
Berlin Zoo	1872–1877	Mare.
BELGIUM		
Antwerp Zoo	1851–1872	Stallion, mare and their two foals – the only quaggas known to have been born in captivity.

Notes

Introduction

1. Smith, *Natural History*, 331.
2. The "gg" in kwagga and quagga is often pronounced as "hh," as noted by Selous: "This name is, however, not pronounced in the English way, 'kwagger,' but kwā-hā, in imitation, doubtless, of the cry emitted by the animal." Selous, "Burchell's Zebra," 81. Pringle stated that the name was "pronounced *quagha* or *quacha*": Pringle, *African Sketches*, 503, and Greig advocated pronouncing the "gg" in quagga as the "ch" in "loch": Greig, "The Origin of the Name 'Quagga'," 149.
3. A social unit of a stallion, one or more mares, and their foals has been variously termed a "band," "breeding group," "family," "group," or "harem." I have chosen to refer to them as a "breeding group."
4. Harris, *Portraits of the Game*, 11; L. Green, *Karoo*, Chapter 3.
5. The range of quaggas is based on Rookmaaker, *Zoological Exploration, 1659–1790*, 287; Skead, *Historical Mammal Incidence, 2*, 568; Boshoff, Landman, and Kerley, "Filling the Gaps on the Maps," 23. Quaggas were said to exist as far east as the Kei River: Bryden, *Kloof and Karroo*, 396. However, the eastern extent of their range north of the Kei River to the Vaal River is not certain. Other maps feature quaggas extending further east within the Orange Free State than shown in this figure, for example, Dolan, "Quagga Commemorated," 15; Greig, "Extinction of the Quagga," 148; Klein and Cruz-Uribe, "Craniometry of the Genus Equus," 82; Rau, "Das Quagga," 40.
6. The people of southern Africa are described in more detail in Chapter 4. The Khoekhoe were herders of cattle, sheep, and goats. They were once known among Europeans by the pejorative name "Hottentot"; I have replaced that term with "Khoekhoe" in all quotations and it is reproduced within brackets in the captions of illustrations. I use the term "Bushmen" rather than "San" for the hunter-gatherers. As a generic term for hunter-gatherers and non-Bantu-speaking herders, I use "Khoe-San." For discussion of the terms "San" and "Bushmen," see Barnard, *Bushmen*.
7. The |Xam (pronounced /ˈkaːm/ in the International Phonetic Alphabet) were hunter-gatherers who have been classified as Bushmen.
8. The term "endling" has been coined to denote the last surviving member of a species.
9. Chadwick, "Saving Stripes."
10. Carruthers, *National Park Science*; Skinner and Chimimba, *Mammals*.

11. In 2013, there were an estimated 2,790 Cape mountain zebras. Hrabar and Kerley, "Conservation Goals."
12. Anon, "Bontebok;" Carruthers, *National Park Science*.
13. Carruthers, *National Park Science*.
14. Anon, "Quagga Project," 1987–2000.

Chapter 1

1. Burchell, *Travels*, vol. 2, 315.
2. The philologist Dr. Wilhelm Bleek recorded, "‖kabba (is a) 'striped quagga' ‖Ʞkhwi̅ (is the 'rechte' [right]) quagga (which is black in the forepart of his back, and white in the hind quarter)." W. H. I. Bleek and Lloyd, "Digital Bleek and Lloyd," BC 151 A2 1 074 5997. See also D. F. Bleek, *Bushman Dictionary*, 773, 746.
3. Feely, "IsiXhosa Names."
4. Somerville, Bradlow, and Bradlow, *William Somerville's Narrative*; Burchell, *Travels*, vol. 1, 420.
5. Feely, "IsiXhosa Names."
6. The Family Equidae, together with rhinoceroses and tapirs, forms the Order Perissodactyla of the Class Mammalia.
7. Cucchi et al., "Detecting Taxonomic and Phylogenetic Signals"; Duncan and Groves, "Genus Equus – Asses, Zebras."
8. MacClintock, *Natural History of Zebras*.
9. McHorse, Biewener, and Pierce, "Fossil Horses."
10. Estes, *Behaviour Guide*.
11. Orlando et al., "Recalibrating Equus Evolution."
12. Information is based on several sources, including Jónsson et al., "Speciation with Gene Flow"; Steiner and Ryder, "Molecular Phylogeny"; Vilstrup et al., "Mitochondrial Phylogenomics."
13. Information in this section is from Estes, *Behaviour Guide*; King and Moehlman, "Equus quagga"; Penzhorn, "*Equus zebra* Mountain Zebra"; Rubenstein et al., "Equus grevyi"; Skinner and Chimimba, *Mammals*.
14. Boddaert, *Sistens Quadrupedia*; Von Linné and Gmelin, *Systema Naturæ*.
15. Groves, "Taxonomy"; Groves and Bell, "New Investigations."
16. Jónsson et al., "Speciation with Gene Flow." Steiner and Ryder also gave a date of about 1.6 million years ago for the divergence of zebra species. Steiner and Ryder, "Molecular Phylogeny."
17. Manes are absent or short in *Equus quagga borensis*, the maneless zebra which occurs at the northern end of the plains zebra's range.
18. Upper and lower jaws each have six incisors, six premolars, and six molars; canines are present in stallions but reduced or absent in mares.
19. Klingel, "Equus quagga Plains Zebra," 433; Estes, *Behaviour Guide*.
20. A. Cobb and S. Cobb, "Do Zebra Stripes Influence Thermoregulation?"
21. Some authors cite a slighter longer gestation period.
22. Rubenstein et al., "Similar but Different."
23. An oxpecker living on a rhinoceros is known to warn its host of an approaching predator and may well do the same for a zebra host: Plotz and Linklater, "Red-Billed Oxpeckers."

24. Estes, *Behaviour Guide*.
25. Estes, *Behaviour Guide*, 240.
26. How et al., "Zebra Stripes, Tabanid Biting Flies."
27. Klingel, "Equus quagga Plains Zebra," 436.
28. Eschner, "Those Little Birds."
29. Information in this section is from Estes, *Behaviour Guide*; Penzhorn, "*Equus zebra* Mountain Zebra;" Penzhorn, "Equus zebra"; Skinner and Chimimba, *Mammals*.
30. Skinner and Chimimba, *Mammals*.
31. Hartmann's mountain zebras occurred in southern Angola and may still be present there; they have also been introduced into South African reserves. Mountain zebras have sometimes been termed "dauw," which echoes their names in isiXhosa, "idawuwa": Feely, "IsiXhosa Names."
32. Estes, *Behaviour Guide*.
33. Penzhorn, "*Equus zebra* Mountain Zebra."
34. Skinner and Chimimba, *Mammals*.
35. Penzhorn, "*Equus zebra* Mountain Zebra."
36. Penzhorn, "*Equus zebra* Mountain Zebra." Although this behavior, termed coprophagy, has not been reported in plains zebras and Grévy's zebras, it probably occurs in these species, too.
37. Hrabar et al., "Equus zebra Ssp. zebra."
38. Gosling et al., "Equus zebra Ssp. hartmannae."
39. Information in this section is from Estes, *Behaviour Guide*; Groves and Bell, "New Investigations"; King and Moehlman, "Equus quagga"; Klingel, "Equus quagga Plains Zebra."
40. Although many taxonomic names are given to honor a person, sometimes just the opposite is intended. This practice extends back to Linnaeus himself, who named a weed *Siegesbeckia* after Johann Siegesbeck who had criticized his sexual descriptions of flowers as "loathsome harlotry." Burchell had quarreled with the staff of the British Museum and so the use of the Latin word "Asinus" for ass (or fool) preceding Burchell's name might have been revenge by Gray, an employee of the Museum. Stewart and Warner, "William John Burchell."
41. The *International Code of Zoological Nomenclature* that formulates the naming of animals uses precedence to determine a correct name: the first valid binomial name of a species is the one to be used.
42. Pedersen et al., "Southern African Origin"; Jónsson et al., "Speciation with Gene Flow."
43. Cabrera quoted Buckley's observations, "Out of five of these animals shot in one herd, there were individuals showing every variation of colour and marking, from the yellow and chocolate stripes, to the pure black and white, the stripes in some ceasing above the hock, and in others being continued distinctly down to the hoof." Cabrera, "Subspecific and Individual Variation," 89; Buckley, "Past and Present Geographical Distribution," 282.
44. Cabrera, "Subspecific and Individual Variation."
45. Groves and Bell, "New Investigations."
46. Rau, "Additions to the Revised List," 41.
47. Lorenzen, Arctander, and Siegismund, "High Variation."
48. Pedersen et al., "Southern African Origin."

49. Smuts, "Pre- and Postnatal Growth;" Sachs, "Live Weights and Body Measurements." Smuts termed the zebras in Kruger National Park *Equus burchelli antiquorum* but this subspecies is now recognized as *Equus quagga burchellii*.
50. Klingel, "Equus quagga Plains Zebra," 434; Estes, *Behaviour Guide*.
51. Skinner and Chimimba, *Mammals*.
52. Smuts, "Pre- and Postnatal Growth"; Smuts, "Reproduction in the Zebra."
53. Klingel, "Equus quagga Plains Zebra."
54. Klingel, "Equus quagga Plains Zebra," 431.
55. Klingel, "Equus quagga Plains Zebra," 433; Estes, *Behaviour Guide*.
56. Neuhaus and Ruckstuhl, "Link between Sexual Dimorphism."
57. Janis, "Evolutionary Strategy of the Equidae." The "complementary feeding" of wildebeest and plains zebras on the same plants works well for both species. Wildebeests are ruminants with a high-quality foraging strategy – eating the leaves which are rich in cytoplasm. By contrast, plains zebras have a high-intake strategy, eating large amounts of fibrous stems containing cellulose which is broken down in their hindguts and lignins which are indigestible.
58. Cain, Owen-Smith, and Macandza, "Costs of Drinking."
59. Cain, Owen-Smith, and Macandza, "Costs of Drinking."
60. Klingel, "Equus quagga Plains Zebra," 431.
61. Fischhoff et al., "Social Relationships."
62. Naidoo et al., "Newly Discovered Wildlife Migration."
63. Harris, *Portraits of the Game*, 29.
64. Hack, East, and Rubenstein, *Status and Action Plan*; King and Moehlman, "Equus quagga."
65. P. Dutton and S. Dutton, "Tragedy in the Making." Some biologists regard Selous zebras as a separate subspecies, *Equus quagga selousi*: Beilfuss et al., "Status and Distribution," 47.
66. Most information in this section is from Estes, *Behaviour Guide*; Williams, "Equus grevyi"; Rubenstein et al., "Equus grevyi."
67. Estes, *Behaviour Guide*; Williams, "Equus grevyi."
68. Williams, "Equus grevyi"; Rubenstein et al., "Equus grevyi."
69. Davidson et al., "Borrowing from Peter."
70. Cordingley et al., "Is the Endangered Grevy's Zebra Threatened by Hybridization?"
71. Anon., "Grevy's Zebra Trust."
72. Burchell, *Travels*, vol. 1, 420.
73. Cuthbert John (Jack) Skead documented this confusion and provided much historical information: Skead, *Historical Mammal Incidence*, 2:563–64.
74. This lack of distinction between different zebras is evident in the frustration of a man who believed he had seen quaggas in South-West Africa in 1932, but found little interest in his observations among his companions as, "to most South-West Africans all zebra were quagga:" Anon., "Quagga Not Extinct."
75. Rau, "Revised List," 42; Skead, *Historical Mammal Incidence*.
76. These explorers, hunters, and naturalists included Thomas Baines, John Barrow, William John Burchell, Samuel Daniell, Robert Gordon, Cornwallis Harris, Henry Lichtenstein, William Somerville, and Anders Sparrman. Their observations will be discussed in later chapters.

Chapter 2

1. From "Afar in the Desert" by Thomas Pringle (1789–1834): Pringle, *African Sketches*, 11, 12. An alternative version is, "Where the timorous Quagga's wild whistling neigh,/ Is heard by the fountain at fall of day . . ." Harris, *Wild Sports*, vi.

2. These hairless patches of skin were also termed "corns," "chestnuts," or "callosities." They are visible in Daniell, *African Scenery and Animals, 1804–1805*, Plate 15; Edwards, "Live Quagga Photographs," 7; Maréchal, "Quagga."

3. Bank, "Anthropology and Portrait Photography."

4. I am grateful to Professor Kevin Dietrich of Stellenbosch University for this information.

5. Barrow, *Travels into the Interior*, vol. 1, 239. In April 2018, an exhibit at the Iziko South Africa Museum in Cape Town gave the date of 80,000 to 100,000 years before the Common Era as the time when humans mixed ocher with fats and organic compounds to make the pigments used in the rock paintings of the Blombos Cave (about 300 kilometers east of Cape Town).

6. Bleek and Lloyd, "Digital Bleek and Lloyd," BC 151 A2 1 074 5997.

7. The name qua-cha is cited in Barrow, *Travels into the Interior*, vol. 1, 93. Similar words – couagga, quacha, qua-cha, quaha, quàha, quahkah, and quakka – are used in other historical accounts. Greig cited various versions of the indigenous name, iqwara, !uaxa, iQwarha, but Feely's more recent account used the word iqwarha in which the "q" represents a clicking sound: Feely, "IsiXhosa Names;" Greig, "Origin of the Name," 149.

8. Thom, *Journal*, vol. 3, 300. Jan van Riebeeck was the founder of the Cape Colony.

9. Sparrman, *Voyage*, vol. 1:223, 224.

10. Gordon, "Second Journey," Diary entry of November 22, 1777.

11. Rau, "Revised List," 58.

12. As shown in Schlawe, *Steppenzebras Von Sudafrika*, 102. The illustrations of Allamand, Buffon, and Schreber are copies and so are not included in Appendix 1.

13. Edwards, *Gleanings of Natural History*, 27.

14. Edwards, *Gleanings of Natural History*, 29.

15. Others have criticized the depiction of the quagga in the foreground, describing it as "rather blocky and stylized": Gall, "Illustrations of the Quagga," 47, and "stocky": Plumb and Shaw, *Zebra*, 86.

16. Harris's Plate 2 and Smith's Plate XXIV also show the white of the belly extending high on the lateral surface, and the dark stripes restricted to the upper part of the lateral surface: Smith, *Natural History*, 330, Plate XXIV.

17. Somerville, Bradlow, and Bradlow, *William Somerville's Narrative*, 36. Somerville confirmed this description, "The quacha approaches nearer the Horse in his form than the [mountain] Zebra does, his head crest and Ears are much like those of a punch made poney, his chest is broad and the limbs much stronger; his tail is that of an ass." Somerville, Bradlow, and Bradlow, *William Somerville's Narrative*, 68–69.

18. Daniell, *African Scenery and Animals, 1804–1805*. Thomas Sutton has suggested that Daniell did not have enough time before leaving for, or returning from, Africa to master engraving to the extent that he could produce the high quality portrayals in *African Scenery and Animals*, and that his brother William, a skilled engraver, did this work: Sutton, *Daniells*. If this were the case, it is possible that William was so misled

by the equine appearance of quaggas that he featured horse-like tails on the animals and gave the one in the middle ground the proportions of a horse. Alternatively, he may have seen a reproduction of Maréchal's (1793) painting and been influenced by that representation. The inaccurate markings on the lateral surface of the quaggas may also owe more to William Daniell than to Samuel Daniell. See also Heywood, "Ways of Seeing Nonhuman Animals."

19. Eeyore was the pessimistic donkey in A. A. Milne's *Winnie-the-Pooh* stories. The quagga in Schouman's painting had been captured by Robert Gordon and given to the House of Orange.

20. Dorst, "Notice Sur Les Spécimens." La Cépède, La Ménagerie du Muséum National D'Histoire Naturelle; "Le Couagga De Louis XVI."

21. Rau's examination of the Paris taxidermy specimen revealed that although the tail skirt was long, it was "most probably not original as fixing on to tail can be seen": Rau, "Revised List," 75. Consequently, the similarity of its length to Maréchal's painting is inconclusive. However, there is no indication that the dock of the taxidermy specimen is anything other than the original. For photographs of taxidermy specimens see: Rau, "Revised List;" Rau, "Additions to the Revised List;" Ridgeway, "Contributions;" Trouessart, "Le Couagga." My observations on quagga tails are also based on examination of the specimens at the following museums in Germany: the Hessisches Landesmuseum Darmstadt, the Naturhistorisches Museum Mainz (where both a mare and stallion are exhibited), the Naturmuseum Senckenberg in Frankfurt am Main, and the Museum Wiesbaden.

22. Hawkins, "Two Quaggas Belonging to Lord Knowsley."

23. Le Fanu, *Catalogue*.

24. Harris, *Portraits of the Game*, 12. A few points about this description bear comment: Harris also recorded the shoulder height as 12–13 hands, that is, 4 feet to 4 feet 4 inches. Harris, *Portraits of the Game*, 13. The length measured, although not defined here, was "from the nose to the point of the tail," the description used when describing a Burchell's subspecies that was also 8 feet 6 inches in length. Harris, *Portraits of the Game*, 30. Harris also stated that the tails of the Burchell's subspecies were 35 inches long: Harris, *Portraits of the Game*, 30; however, quagga taxidermy specimens and images of plains zebras on the Internet show a range of tail lengths.

25. Harris, *Portraits of the Game*, 12. This erroneous information is repeated in Harris, *Wild Sports*, 343.

26. Harris, *Portraits of the Game*, 14. Information on nipples was confirmed by King, "Field Guide;" Klingel, "Equus quagga Plains Zebra."

27. The similarities between Daniell's and Harris's paintings were also noted by Plumb and Shaw, *Zebra*, and suggested plagiarism to Rainier, "Some Quagga Place-Names."

28. In stating that the tails of quaggas extended beyond the hocks, Harris failed to record the variation that existed. Some tails did extend beyond the hocks, such as the Wiesbaden quagga stallion, but this specimen's tail differs from those in Harris's painting in having a distinct dock (personal observation). Other tails ended at the hocks, and some ended before the hocks. Rau measured the tail length of taxidermy specimens and found that these ranged from 290 to 535 mm, averaging 417 mm: Rau, "Revised List." This measurement was from where the tail left the body to its fleshy tip, and did not extend to the end of the brush. The Quagga Project zebra Nina has a long, full tail (see book cover).

29. Bryden, "The Quagga (Equus quagga)," 78.
30. The quotes are from Groves and Bell, "New Investigations," 191; Lydekker, "Note on the Quagga," 429; Bryden, "The Quagga (Equus quagga)," 78.
31. Harris, *Portraits of the Game*, 13. There is general agreement that a shoulder height of 4 feet 10 inches (1,473 mm) marks the cut-off between a pony and a horse, and so the shoulder height that Harris recorded for his "*petit cheval*" indicates that quaggas were the height of some ponies.
32. de Cuvier, *Class Mammalia*, vol. 3, 464.
33. A horse-like tail also features in Harris's description of *Equus burchellii* – then thought to be a separate species, "Ears and tail equine; the latter thirty-five inches long, white and flowing; muzzle black": Harris, *Portraits of the Game*, 30. Harris contrasted this tail with those of mountain zebras which he described as having distinct docks and being 16 inches long: Harris, *Portraits of the Game*, 156.
34. Harris, *Portraits of the Game*, 11. The embedded quote is from de Cuvier, *Class Mammalia*, vol 3, 465.
35. These descriptions are from Ewart, Constable, and Constable, *Guide to the Zebra Hybrids*, 23; Tegetmeier and Sutherland, *Horses, Asses, Zebras*, 63 and 61; Smith, *Natural History*, 330, respectively.
36. Late in the nineteenth century when quaggas were extinct, the same equine qualities were attributed to plains zebras, which were contrasted with mountain zebras, "Burchell's zebra is not only a larger, but, from a utilitarian point of view, a much better-formed animal than the mountain zebra, which may be described as far more asinine in form. It is also more easily broken to harness, and readily becomes a domesticated animal." Tegetmeier and Sutherland, *Horses, Asses, Zebras*, 52.
37. This issue has not been discussed at length previously; however, Dorcas MacClintock, observing that quagga museum specimens had tails like those of plains zebras, noted that some people had described "fully haired" horse-like tails on quaggas: MacClintock, "Professor Marsh's Quagga Mare," 36.
38. Berger, *Ways of Seeing*.
39. Gould, *Wonderful Life*, 276.
40. In the 1970s, Rau measured, photographed and made notes on most of the 23 quagga specimens; he delegated to others the responsibility of measuring the Edinburgh and Kazan specimens. Rau described his measurement of the head–body length as follows: "following contour of dorsal mid-line, except for portion between upper lip and posterior margin of nostrils and middle of rump to extreme posterior margin of buttocks, where the ruler remained straight. From anterior end of muzzle and posterior end of buttock a line to meet ruler in a 90° angle was imagined": Rau, "Revised List," 51. Rau published further findings in Rau, "Additions to the Revised List." Three of the animals were foals, the Milan quagga was described as an "immature female," and the sex of the Bamburg quagga was uncertain. Omitting these five animals, I compared data of shoulder height and head–body length of thirteen adult female specimens and five adult male specimens. Heywood, "Sexual Dimorphism."
41. MacClintock also cited the shoulder heights of these 18 specimens, together with two omitted from my calculations – an immature female and a specimen of indeterminate sex – but she did not distinguish between male and female specimens or provide additional information. MacClintock, "Professor Marsh's Quagga Mare." Neither MacClintock nor Rau commented on sexual dimorphism in body size.

42. Dr. Rebecca Boswell, a clinical psychology intern, provided this analysis of body measurements: females were significantly longer than males (Females: M = 2,294.23 mm, SD = 120.62 mm; Males: M = 2,070.00 mm, SD = 142.13 mm; t(16) = 3.37, p = .004), and marginally taller than males (Females: M = 1,188.85 mm, SD = 46.29 mm; Males: M = 1,140.00 mm, SD = 48.48 mm; t(16) = 1.98, p = .065).

43. Harris, *Portraits of the Game*, 13. Harris also cited a head–body length of 8 feet 6 inches but used a different criterion for this measurement, and so the dimension he recorded cannot be compared with Rau's.

44. Groves and Bell, "New Investigations."

45. Smuts, "Pre- and Postnatal Growth," 80. Smuts's measurements of the total lengths of freshly-killed zebras differ from the head-body lengths that Rau obtained from taxidermy specimens. Rau and Smuts were using different criteria for this measurement and so the dimensions that they recorded cannot be compared.

46. The north–south reduction in striping of plains zebras and increase in their body size is a well-known cline. The magnitude of sexual size dimorphism corresponds to this cline: stallions in the Serengeti were on average 28.7 kilograms heavier than mares, whereas stallions in Kruger National Park (South Africa) were on average 3.1 kilograms lighter than the mares.

47. Harris, *Portraits of the Game*, 12; Baines, *Journal of Residence*, 2:117.

48. Selous, "Burchell's Zebra," 82.

49. Rau, "Revised List," 73. In contrast, the hide of the Paris quagga, preserved as a taxidermy specimen "Le Couagga de Louis XVI," is darker than its painted image and the tail brush is brown. www.mnhn.fr/fr/couagga-de-louis-xvi.

50. Edwards, *Gleanings of Natural History*, 29.

51. Rau, "Revised List," 45; Rau, "Colouration of the Quagga," 136.

52. Nunan, "In Their True Colors."

53. Anon, "Objectives."

54. The difficulty in deciding whether a particular animal was a well-striped quagga or *Equus quagga burchellii* (termed Burchell's subspecies in this book) is evident in contemplating the Mainz quaggas when it was concluded that only the stallion was unambiguously a quagga whereas the more striped mare was either a quagga or *Equus quagga burchellii*. Groves and Bell, "New Investigations." The gradation in striping and coloration between quaggas and Burchell's subspecies is well illustrated by figure 9 of Rau, "Additions to the Revised List," 42, 43.

55. Baines, *Journal of Residence*, 2:152.

56. Groves and Bell, "New Investigations," 191.

57. Boddaert, *Sistens Quadrupedia*, 160.

58. Linné and Gmelin, *Systema Naturæ*, 213.

59. Grubb, "Types and Type Localities."

60. Grubb, "Types and Type Localities;" Gordon, Raper, and Boucher, *Cape Travels*.

61. Groves and Bell, "New Investigations;" Lydekker, *Catalogue*, 19–21; Roberts, *Mammals*, 246.

62. Gray, "Family Equidae."

63. Lydekker, *Catalogue*, 21. Antonius also included these names (except for *E. quagga trouessarti*) in his account of quaggas. Antonius, "Geographical Distribution of Recent Equidæ."

64. Smith, *Natural History*. *Hippotigris isabellinus*, the "Isabella Quagga" is no longer thought to have been a quagga. Grubb, "Types and Type Localities." The yellowish-buff color of the animal, was termed "isabelline."

65. Groves and Bell, "New Investigations."
66. Boshoff, Landman, and Kerley, "Filling the Gaps on the Maps;" Skead, *Historical Mammal Incidence, 2.*
67. Thackeray, "Zebras from Wonderwerk."
68. Boshoff, Landman, and Kerley, "Filling the Gaps on the Maps." "Karoo" is a Khoekhoe word meaning "dry place," but more than 250 million years ago it was a very different habitat with an inland sea containing amphibians and mammal-like reptiles whose fossils are of great interest to paleontologists. Some of these fossils are displayed in the Iziko South African Museum of Cape Town, and the vegetation that supported these animals is represented by a silicified tree trunk standing in front of the museum. Trees like this one growing in swamps lining the shores of the Karoo Sea formed coal and natural gas whose recent extraction, together with uranium, has brought both benefits and problems, and has changed the face of the Karoo in places.
69. Bryden, *Kloof and Karroo*, 24–25.
70. Watkeys, "Soils," 20; Desmet and Cowling, "Climate of the Karoo," 10.
71. Boshoff, Landman, and Kerley, "Filling the Gaps on the Maps;" Desmet and Cowling, "Climate of the Karoo."
72. Boshoff, Landman, and Kerley, "Filling the Gaps on the Maps;" Desmet and Cowling, "Climate of the Karoo;" Siegfried, "Human Impacts."
73. Bryden, *Kloof and Karroo*, 25–26.
74. Pringle, *African Sketches*, 298. Failure of the rains is discussed by Siegfried, "Human Impacts," 240.
75. Interspersed with these dry periods were wetter times; for example, from 16,000 to 14,000 years ago, a cooler climate with more rainfall caused a lake to form east of Kimberley. This interval was part of a longer period (approximately 18,000 to 12,000 years ago) when grasslands expanded and grazing animals, including mountain zebras and quaggas, increased in numbers, though their populations subsequently decreased when climate change reduced grasslands. Faith, "Palaeozoological Insights," 442. There were also warmer periods: from 7,000 to 6,500 years ago temperatures probably exceeded those of the twentieth century. M.E. Meadows and M.K. Watkeys, "Paleoenvironments," 36–37, 40.
76. Lichtenstein, *Travels in Southern Africa*, 2:4.
77. Midgley and van der Heyden, "Form and Function;" Palmer, Novellie, and Lloyd, "Community Patterns."
78. The plant species present at a particular location on the Nama Karoo depend on a variety of factors including soil quality, temperature, availability of water, and the intensity of grazing. In this environment, nutrient-rich soils, heavy grazing, and droughts favor the growth of shrubs over grasses. Palmer, Novellie, and Lloyd, "Community Patterns," 222. Analysis of preserved pollen grains left behind by ancient vegetation of the Nama Karoo shows the back-and-forth shift between shrubs and grasses over the last 10,000 years. Meadows and Watkeys, "Paleoenvironments," 38, and such changes probably also occurred during earlier times.
79. In spite of Bryden's description that some animals of the Nama Karoo "abounded in extravagant profusion," large numbers of quaggas would only have been present when there was sufficient rainfall to replenish pools and to support growth of forage. Bryden, *Kloof and Karroo*, 25; Dean and Milton, "Animal Foraging and Food," 166–76.

80. Dean and Milton, "Animal Foraging and Food;" Milton, Davies, and Kerley, "Population Level Dynamics," 206.
81. L. Green, *Karoo*.
82. Baines, *Journal of Residence*, vol. 2:136.
83. Baines, *Journal of Residence*, vol. 2:88, 117, 139.
84. Klak, Reeves, and Hedderson, "Unmatched Tempo of Evolution."
85. Mucina et al., "Succulent Karoo Biome."
86. Faith, "Palaeozoological Insights," 438. Some large herbivores – including quaggas – lived in the fynbos biome: Radloff et al., "Strontium Isotope Analyses," but in the twenty-first century Cape mountain zebras that fed exclusively on grassy fynbos were found to have a deficient diet, and these zebras preferentially ate grass when living on reserves where both grasses and fynbos were available: Weel et al., "Cape Mountain Zebra;" Smith et al., "Resolving Management Conflicts."
87. Members of the daisy family (Asteraceae) are common in this type of vegetation and one, Renosterbos (*Elytropappus rhinocerotis*), may have given its name to this vegetation type, although the other possible source of the name is a former inhabitant, the black rhinoceros (*Diceros bicornis*).
88. Boshoff and Kerley, "Potential Distributions of the Medium- to Large-Sized Mammals;" Radloff et al., "Strontium Isotope Analyses."
89. Turnbull, "Back from the Dead;" Turner Corporation, "Bontebok Ridge Reserve."
90. Boshoff, Landman, and Kerley, "Filling the Gaps on the Maps." The isolated patches of savanna biome next to the Albany Thickets are not shown in Figure 2.12.
91. Pringle, *Narrative of a Residence*; Feely, "Last Quagga."
92. Pringle, *Narrative of a Residence*, 81; Van Wyk, Novellie, and van Wyk, "Flora of the Zuurberg."
93. Anon, "Stomach Staggers of the Horse," 591.
94. Novellie et al., "Status and Action Plan," 33; Pringle, *Narrative of a Residence*, 81.
95. Hoare et al., "Albany Thicket."
96. Landman and Kerley, "Dietary Shifts."
97. Pringle, *Narrative of a Residence*, 33.
98. Copeland et al., "Strontium Isotope Investigation."
99. Greig, "Taxidermist's Art."
100. Klingel, "Equus quagga Plains Zebra," 431.
101. Faith, "Palaeozoological Insights," 442.
102. Pedersen et al., "Southern African Origin."
103. Gordon, Raper, and Boucher, *Cape Travels*; Harris reported "interminable herds" and "bands of many hundreds." Harris, *Portraits of the Game*, 11; Lichtenstein, *Travels in Southern Africa*, vol. 2:209.
104. Bryden, *Kloof and Karroo*, 402–3.
105. Harris, *Portraits of the Game*, 11.
106. Gordon, "Equus quagga quagga."
107. Gordon, Raper, and Boucher, *Cape Travels*, 95.
108. Edwards, *Gleanings of Natural History*, 29, 30.
109. Pringle, *African Sketches*, 503, 12.

110. Baines, *Journal of Residence*, vol. 2:117. On another occasion, Baines also saw "a mingled herd of bonte [striped] and common, or half-striped quaggas, wildebeestes and blesboks." Baines, *Journal of Residence*, vol. 2:152. Baines's observations are not mentioned in a contrary account: Groves, "Was the Quagga a Species?" that suggested that the ranges of quaggas and plains zebras had not overlapped and proposed that reports that these animals were sympatric (occurred in the same area) in part of the Orange Free State derived from a misreading of Harris's account in Harris, *Portraits of the Game*. A zone of overlap in the Orange Free State between quaggas and other plains zebras is featured in several maps: Greig, "Extinction of the Quagga," 148; Schlawe, *Steppenzebras Von Sudafrika*, 109; Turnbull, "Back from the Dead," 36, and is also described in other accounts such as Rau, "Additions to the Revised List," 40 and Skead, *Historical Mammal Incidence*, 2:564.

111. Harris, *Portraits of the Game*, 11. Two species of gnus occurred in southern Africa: *Connochaetes taurinus*, the blue wildebeest or brindled gnu, occurred north of the Orange River and existed alongside Burchell's subspecies. *Connochaetes gnou*, the black wildebeest, or white-tailed gnu, occurred further south and associated with quaggas: Bryden, *Kloof and Karroo*, 400; Grubb, "Types and Type Localities;" Harris, *Portraits of the Game*, 21. Close to the Vaal River on his return journey to the Cape Colony, Harris observed, "As if by magic the Brindled Gnoo had suddenly given place during the last three days to the common, or white tailed species, and not another specimen occurred during the remainder of our journey." Harris, *Narrative of an Expedition*, 271.

112. Harris, *Narrative of an Expedition*, 302, 303. This behavior of the three species mirrored that north of the Orange River, where ostriches were associated with a different species of gnu and a different subspecies of plains zebras as described in this account: "We soon perceived large herds of Quaggas [Burchell's subspecies] and Brindled Gnoos, which continued to join each other until the whole plain seemed alive. The clatter of their hoofs was perfectly astounding, and I could compare it to nothing, but to the din of a tremendous charge of cavalry, or the rushing of a mighty tempest. I could not estimate the accumulated numbers, at less than fifteen thousand . . . and the long necks of troops of ostriches were also to be seen, towering above the heads of their less gigantic neighbours, and sailing past with astonishing rapidity." Harris, *Narrative of an Expedition*, 74, 75.

113. Burchell, *Travels*, vol. 2:315.

114. Lichtenstein, *Travels in Southern Africa*, vol. 2:209.

115. Janis, "Evolutionary Strategy of the Equidae."

116. Pocock, "Coloration," 357.

117. Lichtenstein, *Travels in Southern Africa*, vol. 2:26.

118. Lichtenstein, *Travels in Southern Africa*, vol. 2:347.

119. Lichtenstein, *Travels in Southern Africa*, vol. 2:50.

120. Gordon, Raper, and Boucher, *Cape Travels*, 111; Sparrman, *Voyage*, vol. 1:225.

121. Estes, *Behaviour Guide*, 237.

122. Harris, *Portraits of the Game*, 13, 14.

123. Sparrman, *Voyage*, vol. 1:131.

124. This observation is based on the grazing times of plains zebras: Estes, *Behaviour Guide*; Klingel, "Equus quagga Plains Zebra," but was probably true for quaggas, too. Lactating mares and breeding group stallions of plains zebras graze for 63 percent of the time during daylight: Neuhaus and Ruckstuhl, "Link between Sexual Dimorphism."

Chapter 3

1. Selous and Roosevelt, *African Nature Notes*, 131.
2. Caro, *Zebra Stripes*; Ruxton, "Striped Coat Coloration."
3. Kipling, *How the Leopard Got His Spots*. Kipling used "Ethiopian" to denote a Black African.
4. Wallace, *Natural Selection*, 368.
5. Gould, "Hopeful Monsters."
6. Darwin, *Descent of Man*.
7. Ruxton, "Striped Coat Coloration."
8. Scott-Samuel et al., "Dazzle Camouflage;" Von Helversen, Schooler, and Czienskowski, "Are Stripes Beneficial?;" Hughes, Troscianko, and Stevens, "Motion Dazzle."
9. Kingdon, "Subgenus Hippotigris," 419.
10. Roberts, *Johannesburg Star*, 7.
11. Scott-Samuel et al., "Dazzle Camouflage;" Von Helversen, Schooler, and Czienskowski, "Are Stripes Beneficial?"; Hughes, Troscianko, and Stevens, "Motion Dazzle."
12. How and Zanker, "Motion Camouflage."
13. Hayward and Kerley, "Prey Preferences;" Klingel, "Equus quagga Plains Zebra," 431.
14. Estes, *Behaviour Guide*.
15. A. Cobb and S. Cobb, "Do Zebra Stripes Influence Thermoregulation?" These authors also pondered whether small-scale air currents produced by stripes might disrupt the flight of biting flies as they attempted to land on zebras.
16. Ruxton, "Striped Coat Coloration."
17. Cobb and Cobb, "Do Zebra Stripes Influence Thermoregulation?"
18. Waage, "Zebra Got Its Stripes."
19. Egri et al., "Polarotactic Tabanids."
20. Scott-Samuel et al., "Dazzle Camouflage;" Von Helversen, Schooler, and Czienskowski, "Are Stripes Beneficial?;" Hughes, Troscianko, and Stevens, "Motion Dazzle."
21. Caro et al., "Function of Zebra Stripes;" Caro and Stankowich, "Concordance on Zebra Stripes;" Caro, *Zebra Stripes*.
22. Ruxton, "Striped Coat Coloration."
23. Klingel, "Equus quagga Plains Zebra," 431.
24. Wallace, *Natural Selection*, 368.
25. Kingdon, "Subgenus *Hippotigris*." The persistence of stripes over the necks and shoulders of quaggas when lost elsewhere on their bodies seems to support the hypothesis that stripes serve as a target area for grooming; Kingdon, however, did not make this argument, nor did he mention the experimental evidence that stripes deter flies.
26. Bard, "Different Zebra Striping Patterns." Similarly, striping was observed to begin when fetuses were 250–270 days old. Smuts, "Pre- and Postnatal Growth," 98.
27. Bard, "Different Zebra Striping Patterns."
28. Rau, "Colouration Abnormalities."
29. Anon, "For Wild Zebra."

30. Bard, "Model."
31. Bard, "Model;" Caro, *Zebra Stripes*.
32. Van Bruggen, "Last Quagga."
33. Klingel, "Equus quagga Plains Zebra," 429; Penzhorn, "Equus zebra," 438; Rau, "Additions to the Revised List," 40; Rau, "Colouration of the Quagga."
34. This range in coat coloration of plains zebras has been described on several occasions: Pocock, "Coloration," 356; Rau, "Revised List," 42, 43; Selous, "Burchell's Zebra," 79, 80.
35. Larison et al., "How the Zebra," 140452.
36. Caro *Zebra Stripes*, 199.
37. Caro *Zebra Stripes*; Caro and Stankowich, "Concordance on Zebra Stripes."
38. Pocock, "Coloration."
39. Rau, "Revised List," 43.
40. Meadows and Watkeys, "Paleoenvironments," 40.
41. Leonard et al., "Loss of Stripes;" Jónsson et al., "Speciation with Gene Flow;" Pedersen et al., "Southern African Origin."
42. Meadows and Watkeys, "Paleoenvironments," 36.
43. Kimura, *Molecular Evolution*.
44. Caccone, "How a Zebra Lost Its Stripes;" Leonard et al., "Loss of Stripes." Clearly, other explanations have to be sought for the abundant striping of mountain zebras living in montane habitats of southern African under environmental conditions similar to those experienced by quaggas.
45. Kipling, *How the Leopard Got His Spots*.
46. Darwin, *Origin of Species*, 167. The hemionus (*Equus hemionus*), also known as the onager, is the Asian wild ass. The ass is the African wild ass, *Equus africanus*.

Chapter 4

1. Bleek and Lloyd, "Digital Bleek and Lloyd," A2_1_65_5247, 5250–52. Dia!kwain was a |Xam who taught Bleek and Lloyd about his culture and language. This story is quoted with permission of Professor Pippa Skotness, Director of the Bleek and Lloyd Collection.
2. Elphick, *Kraal and Castle*.
3. Ross, *Concise History*, 21–27.
4. Ross, *Concise History*, 21–27.
5. Elphick, *Kraal and Castle*.
6. Penn, *Forgotten Frontier*; Adhikari, *South African Genocide*.
7. Malan and Cooke, *Wonderwerk Cave*.
8. Klein, "Preliminary Report."
9. Lichtenstein, *Travels in Southern Africa*, vol. 2:196–99.
10. Lichtenstein, *Travels in Southern Africa*, vol. 2:196.
11. Bleek, Lloyd, and Theal, *Bushman Folklore*, 361.
12. Daniell, *African Scenery and Animals, 1804–1805*, 10.
13. Contrary to Lichtenstein's description in the text, the men in Figure 4.1 are carrying quivers over their right shoulders.
14. Sparrman, *Voyage*, vol. 1:198–201.

15. Borcherds was a member of the Truter–Somerville expedition in 1801 and 1802; his observations were recorded in Somerville, Bradlow, and Bradlow, *William Somerville's Narrative*, 209–10.

16. Sparrman, *Voyage*, vol. 1, plate 2. (1) and (2) show [Khoekhoe] spears. Bushmen made the bow (3), arrows (4–7) and quiver (5). The arrow in (7) is tipped with bone whereas the arrows in (4) and (5) have metal tips. Note the sinews wrapped above the notch that accommodated the bowstring in (6), and the backward-facing quill behind the arrowhead in (4) and (7). The plate's caption reflects Sparrman's belief that Bushmen were a constituent group of the Khoekhoe.

17. Gordon, Raper, and Boucher, *Cape Travels*, 270.

18. Somerville, Bradlow, and Bradlow, *William Somerville's Narrative*, 210.

19. Lichtenstein, *Travels in Southern Africa*, vol. 2:199.

20. Sparrman, *Voyage*, vol. 1:198.

21. MacKenzie, *Empire of Nature*, 64.

22. Hewitt, *Structure, Meaning and Ritual*.

23. Barrow, *Travels into the Interior*, vol. 1:187.

24. Bleek, Lloyd, and Theal, *Bushman Folklore*, 285–87.

25. Somerville, Bradlow, and Bradlow, *William Somerville's Narrative*, 210; Gordon, Raper, and Boucher, *Cape Travels*, 127.

26. Somerville, Bradlow, and Bradlow, *William Somerville's Narrative*, 210; Gordon, Raper, and Boucher, *Cape Travels*, 357.

27. Harris, *Portraits of the Game*, 13.

28. Burchell, *Travels*, vol. 2:83; Gordon, "Equus quagga quagga;" Harris, *Wild Sports*; Somerville, Bradlow, and Bradlow, *William Somerville's Narrative*, 69.

29. Sparrman, *Voyage*, vol. 1:193–94. Sparrman stands out as a keen observer who was sympathetic to indigenous people and ready to learn from them.

30. Sparrman noted that, "with good horses the people here are used to hunt down the zebras with ease: but who knows, whether both [mountain] zebras and quaggas would not become quicker in their paces by frequent riding and exercise." Sparrman, *Voyage*, vol. 1:226.

31. On the various rifles used in South Africa: Lategan and Potgieter, *Die Boer se Roer*.

32. Storey, *Guns, Race, and Power*, 41.

33. Hudson, "Infantry Weapons."

34. Ross, *Concise History*, 35.

35. MacKenzie, *Empire of Nature*, 87.

36. MacKenzie, *Empire of Nature*, 303.

37. Baker, "William Cotton Oswell."

38. Daniell, *African Scenery and Animals, 1804–1805*, 34.

39. Swart, "Riding High," 18.

40. Hudson, "Infantry Weapons."

41. Storey, *Guns, Race, and Power*, 139.

42. Storey, *Guns, Race, and Power*, 87.

43. Storey, *Guns, Race, and Power*, 132.

44. Bryden, "The Quagga (Equus quagga)," 76; Gordon, "Equus quagga quagga."

45. Selous reported that removal of the fat rendered zebra steaks edible. Selous, "Burchell's Zebra," 82. In 2014, meat from South African plains zebras was sold in Britain for human consumption and was praised as having only a tenth of the fat in a comparable beef steak. The zebra meat was said to have a delicate flavor of game and to taste sweeter than beef: Emma Powell, "Zebra Meat."

46. Gordon, "Equus quagga quagga;" Selous, "Burchell's Zebra," 82.

47. Gordon, Raper, and Boucher, *Cape Travels*, 273; Harris, *Wild Sports*, 258.

48. Simmonds, *Animal Products*, 326; Baines, *Journal of Residence*, vol. 2:149.

49. Gordon, "Equus quagga quagga," Diary entry of November 23, 1777.

50. Somerville, Bradlow, and Bradlow, *William Somerville's Narrative*, 69.

51. Burchell, *Travels*, vol. 2:83.

52. Harris, *Portraits of the Game*, 13.

53. Skead, *Historical Mammal Incidence*, 1:352. Baines, who used the term "riems" for strips of raw hides, wrote that they were "used for purposes as multifarious as those of rope yarns at sea" and illustrated the twisting process by which they were softened: Baines, *Journal of Residence*, vol. 1:214.

54. Buckley, "Past and Present Geographical Distribution," 281.

55. Pringle, *Narrative of a Residence*, 55.

56. Turnbull, "Back from the Dead," 36.

57. Edwards, *Gleanings of Natural History*, 30.

58. Sparrman, *Voyage*, vol.1:224.

59. Anon, "Quagga," *The Victorian Naturalist*, 16.

60. Cited by Skead, *Historical Mammal Incidence*, 1:316.

61. Thom, *Journal*, 3:300. Pieter Meerhoff had made a determined attempt to capture this animal by shooting it and then jumping astride the animal to cut a heel sinew, but it rose from beneath him, kicked him, and escaped.

62. Thom, *Journal*, 1:307.

63. Thom, *Journal*, 3:300.

64. Swart, *Horses, Humans and History*.

65. Swart, *Horses, Humans and History*.

66. Daniell, *African Scenery and Animals, 1804–1805*, 42. Pocock, "Cape Colony Quaggas;" Skead, *Historical Mammal Incidence*; Tegetmeier and Sutherland, *Horses, Asses, Zebras*, 59.

67. Lichtenstein, *Travels in Southern Africa*, vol. 2:212.

68. Harris, *Wild Sports*, 221.

69. Gordon, Raper, and Boucher, *Cape Travels*, 95; Baines, *Journal of Residence*, vol. 2:150.

70. Selous, "Burchell's Zebra," 82.

71. This behavior of quagga foals differs from that of adult plains zebras which flee approaching humans. Brubaker and Coss, "Evolutionary Constraints on Equid Domestication."

72. Skead, *Historical Mammal Incidence*, 1:350.

73. Diamond, *Guns, Germs, and Steel*, 170.

74. Pringle, *Narrative of a Residence*, 142; Pringle, *African Sketches*, 274.

75. Lichtenstein, *Travels in Southern Africa*, vol. 2:204.

76. Harris, *Portraits of the Game*, 14.

77. Gordon, Raper, and Boucher, *Cape Travels*, 111; Sparrman, *Voyage*, vol. 1:225. Even today, some farmers use plains zebras to deter predation by dogs and jackals: Stears, Shrader, and Castley, "Equus quagga." Other equines are also used to defend domestic animals: gelded male donkeys (jacks) and female donkeys (jennies) are used to protect animals against predators such as coyotes, dogs, and foxes. Kept within the same enclosure as calves, goats, and sheep, guardian donkeys bray at predators, which may both repel them and alert humans, and will charge, bite, and kick intruders – just as quaggas did: Dohner, "Donkeys."

78 Diamond, *Guns, Germs, and Steel*, 171.

79. Mentzel, *Cape of Good Hope*, 235.

80. Gordon, "Equus quagga quagga." Diary entry of November 23, 1777.

81. Sparrman, *Voyage*, vol. 1:224–25.

82. Barrow, *Travels into the Interior*, vol. 1:93.

83. Sparrman, *Voyage*, vol. 1:224–25.

84. Tegetmeier and Sutherland, *Horses, Asses, Zebras*, 54.

85. In this and similar instances – such as Lord Rothschild's carriage that was drawn by three zebras and a horse – it might be argued that plains zebras needed to work alongside horses or donkeys. However, evidence against this assertion is provided by the team of four plains zebras that pulled a two-wheeled cart: Tegetmeier and Sutherland, *Horses, Asses, Zebras*, plate opposite 57.

86. Tegetmeier and Sutherland, *Horses, Asses, Zebras*, 55. These authors also addressed the possible use of mountain zebras which they said were, "much wilder and more intractable" than plains zebras: Tegetmeier and Sutherland, *Horses, Asses, Zebras*, 40–41. However, mountain zebras lassoed near George in the Cape Colony were exported to Mauritius, where they were trained to harness: Harris, *Portraits of the Game*, 154. A photograph of a woman sitting sidesaddle on a mountain zebra described as "broken to saddle" is unconvincing evidence that they could be ridden: Tegetmeier and Sutherland, *Horses, Asses, Zebras*, 41.

87. Barnard, "African Horse Sickness," 16. Although it was generally believed that quaggas were resistant to diseases that affected imported equines, Baines recorded that ". . . even the wild quaggas [were] dying of the horse sickness". Baines, *Journal of Residence*, vol. 2:52. He might have misidentified the disease, but it is also possible that quaggas differed from other plains zebras in their susceptibility to African horse sickness.

88. Barrow, *Travels into the Interior*, vol. 1:93.

89. Somerville, Bradlow, and Bradlow, *William Somerville's Narrative*, 36.

90. Lydekker, *Game Animals*, 60.

91. Ewart, Constable, and Constable, *Guide to the Zebra hybrids*, 19, 20.

92. Plumb, "Queen's Ass."

93. Lichtenstein, *Travels in Southern Africa*, 2:162. Lichtenstein noted that this quagga, "suffered himself readily to be stroked and caressed by the people about."

94. Heck, *Animal Safari*, 85.

95. Ewart, Constable, and Constable, *Guide to the Zebra Hybrids*, 39.

96. Skinner and Chimimba, *Mammals*.

97. Skead, *Historical Mammal Incidence*, 1:352. Skead's evaluation, however, overlooks the value that quagga hides acquired in the mid-nineteenth century.

98. Heck, *Animals, My Adventure*, 170.

99. Kellert, *Value of Life*, 38.

100. Pringle, *Narrative of a Residence*, 125.

101. Pringle, *Narrative of a Residence*, 149. Pringle, *African Sketches*, 274. The assessment of hides as "almost useless" predated the value that they acquired in the mid-nineteenth century when they were exported to make expensive boots.

Chapter 5

1. Pearson and Galton, *Francis Galton*, vol. 2:159. Galton was writing to his cousin, Charles Darwin.

2. Most quaggas were exported to Europe, although in the nineteenth century a few were exhibited at circuses in the United States; however, there do not appear to be records of them in zoos there. Barnaby, *Quaggas*, 96.

3. This was true of people as well as animals. The unfortunate Khoekhoe woman Sara Baartman is the most disturbing example: Crais and Scully, *Sara Baartman and the Hottentot Venus*.

4. Cowie, *Exhibiting Animals in Nineteenth-Century Britain*; Grigson, *Menagerie*.

5. Grigson, *Menagerie*; Simons, *The Tiger*; Plumb and Shaw, *Zebra*.

6. Simons, *The Tiger*.

7. Grigson, *Menagerie*.

8. Dorst, "Notice Sur Les Spécimens."

9. Edwards, "Live Quagga Photographs;" Huber, "Documentation of the Five Known Photographs."

10. Barnaby, *Quaggas*, 98, 99. The stallion is illustrated in Figure 0.1 and his death features in Enright's poem in chapter 7.

11. de Cuvier, *Class Mammalia*, vol. 3:465.

12. Barnaby, *Quaggas*, 94; MacClintock, "Professor Marsh's Quagga Mare."

13. Smith, *Natural History*, 331.

14. Edwards, *Gleanings of Natural History*, 30.

15. Gray, "Family Equidae," 247. Gray had a similar assessment of the mountain zebra, "The beautiful female that is in Exeter Change appears to be vicious, for she is fond of being taken notice of, but gradually sidles round, and attempts to kick at her fondler." Gray, "Family Equidae," 248.

16. Burkhardt, "Closing the Door on Lord Morton's Mare;" Ritvo, *Platypus*.

17. Morton, "Communication," 20.

18. Morton, "Communication."

19. Morton, "Communication," 20.

20. Morton, "Communication," 22.

21. Le Fanu, *Catalogue*. The young foal is not mentioned in Morton's account and was presumably born after Morton wrote his report to the Royal Society.

22. Le Fanu, *Catalogue*.

23. Morton, "Communication."

24. Although Weismann coined the term telegony, his germ-plasm theory published in 1893 stated that somatic cells were separate from germ cells and so would not influence the germ line in the way proposed by supporters of telegony. He discussed Lord Morton's observations in a chapter headed, "Doubtful phenomena of heredity:" Weismann, *Germ-Plasm*.

25. Darwin and Darwin, *Foundation*, 108.

26. Darwin, *Variation of Animals and Plants*, vol. 2:374.

27. Darwin, *Variation of Animals and Plants*, vol. 1:404.

28. Flint, *Textbook*, 797. Examples of telegony – mainly in dogs and humans – were also cited in Gould and Pyle, *Anomalies and Curiosities*.

29. Burkhardt, "Closing the Door on Lord Morton's Mare;" Heywood, "Quagga and Science;" Ritvo, *Noble Cows.*

30. Swart, *Horses, Humans and History*, 67.

31. Strindberg, *Father*, Act 2.

32. Pearson and Galton, *Francis Galton*, vol. 2:159.

33. Zola, *Madeleine Férat*, 201.

34. Ewart, Constable, and Constable, *Guide to the Zebra Hybrids*, 44.

35. Ritvo, *Noble Cows.*

36. Ewart, Constable, and Constable, *Guide to the Zebra Hybrids.*

37. Ewart, *Penycuik Experiments*, 137.

38. Darwin, *Origin of Species*, 163–67.

39. Lord Morton's quagga and its influence on ideas of heredity have been largely forgotten, but belief in telegony continues in the twenty-first century. The idea that the first mate can affect offspring conceived by a later partner has been used as an argument for pre-marital chastity. In 2008, I read an Internet posting in which the owner of a Boston Terrier that had been impregnated by another breed of dogs was advised, "now she cannot breed with another Boston, due to the contamination of the blood line." Interestingly, there are now reports that an earlier mate *can* influence the offspring conceived by a later mate; however, the effect is not genetic, being brought about by seminal fluid: Crean, Kopps, and Bonduriansky, "Revisiting Telegony;" Simmons and Lovegrove, "Nongenetic Paternal Effects."

40. The misattribution of stripes to the effects of Lord Morton's quagga parallels the situation described in Chapter 2 where conceptions that quaggas resembled horses led to descriptions and depictions of their having horse-like tails.

41. Barnaby, *Quaggas.*

42. Gens, *Promenade Au Jardin Zoologique.*

Chapter 6

1. Bryden, *Kloof and Karroo*, 402.

2. Sclater, *Guide to the Gardens*, 23.

3. Greig, "Taxidermist's Art," 140.

4. Skinner, "Further Light."

5. Beinart, "The Night of the Jackal;" Rau, "Revised List," 41.

6. Gordon, "Equus quagga quagga."

7. Burchell, *Travels*, vol. 2:81–82.

8. Barrow, *Travels into the Interior*, vol. 2:357; Lichtenstein, *Travels in Southern Africa*, vol. 2:74.

9. Cape of Good Hope, "Proclamation," 150.

10. Cape of Good Hope, "Proclamation," 155, 151; MacKenzie, *Empire of Nature.*

11. Pringle, *Narrative of a Residence*, 109; Bryden, "The Quagga (Equus quagga)," 77.

12. Baines, *Journal of Residence*, vol. 2:139.

13. For other histories of the progressive extermination of quaggas, see Cary [Mungall], "History and Extinction of the Quagga;" Mungall [née Cary], "Extinction."

14. Two quagga foals born at the Antwerp Zoo could have been the beginning of a captive breeding program, but apparently no further animals were born: Gens, *Promenade Au Jardin Zoologique.*

15. Skead, *Historical Mammal Incidence*, 2:565; Bryden, "The Quagga (Equus quagga)," 72. Neither author cited a primary source.

16. Greig, "Lesson of the Quagga," 134.

17. Berridge and Westell, *Book of the Zoo*; Protheroe, *New Illustrated Natural History.*

18. Cited by MacKenzie, *Empire of Nature*, 206. Sir John Kirk had been Consul General of Zanzibar.

19. Barnett et al., "Lost Populations."

20. Bryden, "The Quagga (Equus quagga)," 76. Bryden's characterization is at odds with his own description of quaggas as being "fleet": Bryden, *Kloof and Karroo*, 403, and with Harris's observation that, "Judging from its low, and somewhat laboured pace, the inactive spectator would pronounce the Quagga to be a slow and heavy galloper; but it is only necessary to follow its flight a few yards on horseback, to be convinced of the rate at which it covers the ground." Harris, *Portraits of the Game*, 13, 14.

21. Somerville, Bradlow, and Bradlow, *William Somerville's Narrative*; Sparrman, *Voyage*, vol. 1:226.

22. Beddard, *Natural History*, 61.

23. Eloff, "Passing of the True Quagga." Arguments about adaptation to the environment were a theme in Eloff's writings, for example, Eloff listed the physical characteristics of "Boers" that enabled them to flourish in the South African environment. Dubow, "Afrikaner Nationalism."

24. Paddle, *Last Tasmanian Tiger*; Freeman, "Imaging Extinction."

25. Greenberg, *Feathered River.*

26. Hoffman et al., "Historical and Contemporary Land Use." It is likely that recent decrease in grasses was and continues to be caused by increasing levels of atmospheric CO_2 (sometimes termed "CO_2 fertilization") due to burning fossil fuels. Some grasses have a type of photosynthesis that enables them to take up CO_2 through stomata (pores) on their leaves without losing excessive amounts of water by transpiration. These grasses belong to a group of plants termed "C4 plants," and are often able to survive arid conditions. Trees and shrubs, by contrast, are termed "C3" plants and often transpire large amounts of water through their stomata in order to take up CO_2 into their leaves; this is also true for C3 grasses that occur in less arid areas than C4 grasses. Consequently, higher atmospheric CO_2 levels benefit C3 plants, but not C4 plants. CO_2 fertilization means that woody plants have an increasing advantage over C4 grasses in dry areas such as the Karoo and are gradually replacing them. For further discussion of replacement of grasses by woody plants see Jacobs, *Environment, Power, and Justice*, chapter 7.

27. Beinart, "The Night of the Jackal;" Hoffman et al., "Historical and Contemporary Land Use."

28. Milton, Davies, and Kerley, "Population Level Dynamics."

29. Beinart, "The Night of the Jackal."

30. Jackson, *Manna in the Desert*, 73. Others have also pondered whether the drought that began in 1877 contributed to extinction. MacClintock, "Professor Marsh's Quagga Mare;" Rau, "Revised List," 42.

31. Rau, "Revised List," 42.

32. Rau, "Revised List." The 1822 Proclamation listed bonteboks, elephants, and hippopotami as animals whose killing required special permission: Cape of Good Hope, "Proclamation," 152.

33. Skinner, "Nowhere to Run."

34. Bryden, "Mountain Zebra (Equus zebra)," 98; Kingdon, *Kingdon Field Guide*, 469.

35. Chadwick, "Saving Stripes."

36. Anon, "Jan Christoffel Greyling Kemp (1872–1946)."

37. By 1950, only two stallions were left in the park, but five stallions and six mares that had been saved on farms were moved there and the population has increased steadily since then: Skinner and Chimimba, *Mammals*.

38. Harris, "Extinct Kin of Zebras," 8.

39. Turnbull, "Back from the Dead," 37.

40. Ross, *Concise History*, 35–53.

41. Van der Merwe, *Migrant Farmer*, 32.

42. Beinart, "The Night of the Jackal;" Mungall, "Extinction."

43. Carruthers, "Changing Perspectives on Wildlife;" Van Sittert, "Bringing in the Wild."

44. Bryden, "Blesbok," 189.

45. Gordon-Cumming, *Hunter's Life*, 68.

46. MacKenzie, *Empire of Nature*.

47. Ritvo, "Q Is for Quagga;" Harris, *Narrative of an Expedition*, 14.

48. Harris, *Narrative of an Expedition*, xi.

49. Harris, *Wild Sports*, 283. Harris had the sable antelope skin mounted and provided a description of the species and a binomial name. Harris, "New Species of Antelope;" Harris, "*Aigocerus niger.*"

50. Harris, *Wild Sports*, 284. Harris's books were: *Narrative of an Expedition into Southern Africa, during the Years 1836, and 1837* which was published in 1838, and was republished in a revised form in 1839 as *The Wild Sports of Southern Africa*, and *Portraits of the Game & Wild Animals of Southern Africa*, published in 1840. Harris's "shooting madness" seems pathological, but he provided some useful scientific observations in his books. Inaccuracies are also present, for example, as described in chapter 2 he incorrectly stated that quaggas and plains zebras (then viewed as separate species) had four nipples and full horse-like tails.

51. Clarke, *Overkill: The Race to Save Africa's Wildlife*; MacKenzie, *Empire of Nature*, 115.

52. Baines, *The Greatest Hunt*.

53. Watson and Clifford, "Mapping Supply Chains."

54. MacKenzie, *Empire of Nature*.

55. Isenberg, *Destruction of the Bison*, 129.

56. The Spectator, "Man the Destroyer." Quagga hides were also used as connecting bands for machinery. Buckley, "Past and Present Geographical Distribution."

57. Anon, "Exit Quagga," 100.

58. The Spectator, "African Skins," 17.

59. Ross, *Concise History*, 54–57.

60. Gatti, *Here Is the Veld*, 36. See also Ellis, *South African Sketches*, 216–17.

61. There is an unfortunate parallel between quaggas becoming a valuable commodity in the nineteenth century and succulent plants assuming this role currently:

https://usanewslab.com/world/in-south-africa-poachers-now-traffic-in-tiny-succu
lent-plants/. Large-scale theft of these plants from South African deserts supplies
collectors throughout the world and leads to loss of biodiversity and possible extinction in their native land.

62. Bryden, *Kloof and Karroo*, 401. Bryden's timing seems incorrect as "the past twenty
years" would have been 1869–89, and quaggas were probably already extinct in the
wild during the latter half of this period. The period of 1840–75 was when wild
animals were extirpated from the Orange Free State, with the pace increasing as the
hide trade developed in the 1850s: MacKenzie, *Empire of Nature*, 115.

63. Anon, "Laws Too Late."

64. Anon, "Exit Quagga," 100.

65. The Spectator, "Man the Destroyer," 10.

66. MacKenzie, *Empire of Nature*; Van der Merwe, *Migrant Farmer*; Van Sittert, "Bringing
in the Wild."

67. Cloudsley-Thompson, "Frederick Courteney Selous."

68. Anon, "Laws Too Late."

69. Cape of Good Hope, "Act for the Better Preservation of Game," 1886 No. 36, cited
in Greig, "Lesson of the Quagga," 135.

70. Cape of Good Hope, "Proclamation."

71. This situation could be reversed in the future, with domestic equines facing increased
exposure to African horse sickness as zebras are introduced (or reintroduced) to some
areas when used to stock game reserves and wildlife areas: Barnard, "African Horse
Sickness."

72. Hopkins, "Binsey Poplars."

73. Groves, "Was the Quagga a Species?"

74. Bryden, *Kloof and Karroo*, 402–3.

75. Selous and Roosevelt, *African Nature Notes*, 131. Selous made similar comments
about the extinction of blueboks (*Hippotragus leucophaeus*), which he viewed as being
comparable in appearance to small roan antelopes.

76. Bales, *Ghost Birds*.

77. Greenberg, *Feathered River*.

78. Jarvis, "Paper Tiger."

79. Fitzsimons, *Natural History*, 3:180.

80. Rose, *Bushman, Whale*, 64.

81. Anon, "Laws Too Late."

82. Anon, "Quagga Not Extinct."

83. To explain why a fully-striped mountain zebra might look like a quagga, Roberts
postulated that "In certain positions the body stripes may not show up, while those
of the neck show up conspicuously, or the reverse may happen – all depending upon
the light and position of the animal." Roberts, *Johannesburg Star*, 7. This explanation,
however, seems unlikely as studies on the luminance of white and black stripes of
Grévy's zebras showed that a contrast between them existed under a variety of
lighting conditions. Caro, *Zebra Stripes*, 44. Incorrect identification of mountain
zebras was not the only possibility: two observers with field experience in South-
West Africa suggested that occasionally occurring hybrids between zebras and horses
or asses might have been misidentified as quaggas. Carp, "A Quagga Chase in South-

West Africa;" Heck, *Animal Safari*, 85. Yet, it is highly unlikely that as many as fourteen hybrids would exist and would herd together.

84. Hahn's report was said to state that the bright noontime sun made the stripes difficult to see and resulted in these animals resembling dark brown donkeys. Anon, "The Quagga," (August 1951). The notion that the black and white stripes of mountain zebras could be seen as the brown color of quaggas is not supported by Caro, *Zebra Stripes*. Wishful thinking influenced people in believing that they had seen quaggas.

85. Anon, "Quagga Not Extinct."

86. Keynes, *Quentin Keynes*. Keynes and his companion did film the remains of the *Dunedin Star*, a British merchant ship that famously wrecked on the Skeleton Coast in 1942.

87. Büttiker-Otto, *Memories of a Scientist*; Carp, *I Chose Africa*.

88. Carp, "Quagga Chase in South-West Africa.".

89. Boyle, C. L., "The Quagga."

90. Allen, D. S. E. "The Quagga."

91. Anon, "To Search for Quagga."

92. Carruthers, *National Park Science*.

93. Thamm, "Rebirth of the Quagga."

94. Rau, "Revised List," 57.

95. Gens, *Promenade Au Jardin Zoologique*.

Chapter 7

1. Anon, "To Search for Quagga."

2. Adhikari, *South African Genocide*.

3. De Prada-Samper, "Forgotten Killing Fields," 177.

4. Bank, *Bushmen in a Victorian World*. Decades later, Lloyd published a selection of the stories. Bleek, Lloyd, and Theal, *Bushman Folklore*.

5. Bleek and Lloyd, "Digital Bleek and Lloyd," A2_1_74_05993–97.

6. Bleek and Lloyd, "Digital Bleek and Lloyd," A2_1_74_05995–6002.

7. Bleek and Lloyd, "Digital Bleek and Lloyd," A2_1_52_04073.

8. Bleek and Lloyd, "Digital Bleek and Lloyd," A2_1_72_5881–83.

9. Bleek and Lloyd, "Digital Bleek and Lloyd," A2_1_74_05993–96. In commenting on this action, Dia!kwain's father observed that quagga's meat was superior to that of other game. This opinion echoed Somerville's report that the Khoekhoe esteemed quagga meat as "the most delicate of any that the country produces." Somerville, Bradlow, and Bradlow, *William Somerville's Narrative*, 69.

10. Bleek and Lloyd, "Digital Bleek and Lloyd," A2_1_50_03898–915.

11. Bleek and Lloyd, "Digital Bleek and Lloyd," A2_1_104_08603–14. This story is quoted with permission of Professor Pippa Skotness, Director of the Bleek and Lloyd Collection.

12. Bleek and Lloyd, "Digital Bleek and Lloyd," A2_1_87_07114–56.

13. Bleek and Lloyd, "Digital Bleek and Lloyd," A2_1_9_00466.

14. Bleek and Lloyd, "Digital Bleek and Lloyd," A2_1_68_05457–77.

15. Bleek and Lloyd, "Digital Bleek and Lloyd," MP2_109.

16. Bleek and Lloyd, "Digital Bleek and Lloyd," A2_1_65_05246–57. The epigraph of Chapter 4 includes part of this story.

17. Bleek and Lloyd, "Digital Bleek and Lloyd," A2_1_57_04574–617.

18. Bleek and Lloyd, "Digital Bleek and Lloyd," A2_1_105_08651–58.

19. Hewitt, *Structure, Meaning and Ritual*.

20. On the history of apartheid, see Ross, *Concise History*, 114–62.

21. De Prada-Samper and de Villiers, *Man Who Cursed the Wind*.

22. De Prada-Samper, personal communication to the author, October 1, 2020.

23. Jacobs, "Great Bophuthatswana Donkey Massacre."

24. De Prada-Samper, "Roads, Ghosts and Rock Paintings." De Prada-Samper finds evidence for both ghost wagons and bones in Ellis, *South African Sketches*, 216–17.

25. Anon, "Harvard-MIT Mathematics Tournament," 5.

26. Mills et al., "Quagga Mussel."

27. Limburg et al., "The Good, the Bad, and the Algae."

28. Carpenter, "Halaelurus quagga (Alcock, 1899), Quagga Catshark."

29. Witz, "Huberta's Journey."

30. Skead, *Historical Mammal Incidence*, 2:564.

31. Heck, *Animal Safari*, 86.

32. Heck featured in the Hollywood adaptation of *The Zookeeper's Wife* set at the Warsaw Zoo during World War II: Caro, "Zookeeper's Wife" (Focus Features, 2017).

33. Skead, *Historical Mammal Incidence*, 2:566–67.

34. Pringle, *African Sketches*, 11, 12, 75, 76. *Afar in the Desert* is cited frequently but not *The Desolate Valley*, part of which reads: "Round this secluded region circling rise/A billowy waste of mountains, wild and wide;/Upon whose grassy slopes the pilgrim spies/The gnu and quagga, by the greenwood side,/Tossing their shaggy manes in tameless pride . . ."

35. "The Extermination of the Quagga" is in the Iziko Museums of South Africa Art Collections in Cape Town.

36. Ford, Katz, and Kazanjian, *Tigers of Wrath*, 15.

37. Burns, "Quagga Stamps."

38. Dolan, "Quagga Commemorated;" Barnaby, *Quaggas*, 69.

39. Gatti, *Here Is the Veld*.

40. Macnab, "Winged Quagga."

41. Macnab, "The Quagga's World."

42. Lindsay, "Elegy for the Quagga," from *Twigs & Knucklebones*. Copyright © 2008 by Sarah Lindsay. Reprinted with the permission of The Permissions Company, LLC on behalf of Copper Canyon Press, coppercanyonpress.org.

43. Enright, "The Quagga," From *Selected Poems* by D. J. Enright. D. J. Enright is the copyright holder and Watson, Little Ltd is the licensing agent. Reprinted with permission of Watson Little Ltd.

44. Green, *Quagga*.

45. Wende et al., *Quagga's Secret*.

46. Bradfield, *All That Glitters*.

47. Morrison and Watson, *Seven Moon Circus*.

48. Silverberg, *Born with the Dead*; Crichton, *Jurassic Park: A Novel*.

49. Barnaby, *Quaggas*; Barnaby, Adams, and Bartlett, *Quagga Quotations*; Spreen, *Monument* ; Cary [Mungall], "History and Extinction of the Quagga."

50. MacClintock, *Natural History of Zebras*; Plumb and Shaw, *Zebra*.

51. Ritvo, *Noble Cows*; Ritvo, "Q Is for QUAGGA;" Swart, "Frankenzebra;" Swart, "Zombie Zoology;" Swart, "Resurrection Conservation."

52. Silverston, "Khumba;" Govindasamy, "Stars and Stripes."

53. Gatti, *Here Is the Veld*.

54. The writings of two quagga enthusiasts, David Barnaby and Reinhold Rau, provided much of the background information on museum specimens: Barnaby, *Quaggas*; Rau, "Revised List;" Rau, "Additions to the Revised List."

55. Rau, "Revised List;" Rau, "Additions to the Revised List."

56. Rau, "Revised List," 50.

57. Larkin and Porro, "Three Legs Good."

58. MacClintock, "Professor Marsh's Quagga Mare."

59. Barnaby, *Quaggas*; Rau, "Revised List."

60. Most museum specimens were prepared by mounting a quagga skin on a framework, termed a manikin. The process began with taxidermists observing the animal and measuring it; this was not always possible as sometimes their hides were shipped from Africa. Taxidermists detached the skin complete with hooves, removed any remaining flesh, and treated the skin with preservatives. Next, they constructed a manikin whose foundation was often the cleaned skeleton of the animal; however, if the skeleton was also to be exhibited, taxidermists constructed the manikin using casts of the bones, the skeletons of other equines, or other material. They positioned the cleaned skeleton supported by iron bars in the desired stance and covered it with wire netting to which papier-mâché was applied until the manikin recreated the shape of the animal's body. They worked the skin into place on the manikin – removing or adding papier-mâché to achieve the best fit – and sewed it in such a way that the stitches were not seen. Taxidermists completed their work by inserting glass eyes and, if the ear cartilage had been removed, they stiffened and positioned the ears using cardboard or papier-mâché: Rose, *Bushman, Whale*, 40–41; Rowley, *Art of Taxidermy*, 146–58.

61. Rau, "Revised List;" Rau, "Additions to the Revised List;" Rau, "Quaggas in Museums."

62. Denise Hamerton, Curator of Terrestrial Vertebrates at the Iziko South African Museum in Cape Town, interview with the author, April 24, 2018.

63. March Turnbull, personal communication to the author, April 19, 2018; Hulley, "Quagga Pioneer."

64. Hulley, "Quagga Pioneer."

65. Rau, *Rough Road*, 1.

66. From 1907 to 1923, nine expeditions from the South African Museum made photographs, measurements, and whole-body molds of living Bushmen. Bribed with commodities such as tobacco, Bushmen were required to assume poses for museum dioramas and were then encased in plaster of Paris: Rose, *Bushman, Whale*, 44–50. The experience must have been traumatic, especially as the setting plaster emitted heat, "The strain was awful ... One feels as if one were being suffocated, especially when the mould of the face is being taken," reported the bodybuilder Eugen

Sandow, who experienced this technique in a different setting: Josh Davis, "Eugen Sandow."

67. Rau, "Bushmen."
68. De Prada-Samper, "Forgotten Killing Fields."
69. Baard, "Geometric Tortoise;" Hulley, "Quagga Pioneer."
70. Max, "Can You Revive an Extinct Animal?"
71. Rau, "Revised List;" Rau, "Additions to the Revised List."
72. Rau, "Revised List," 80.
73. Rau, "Revised List;" Rau, "Additions to the Revised List."
74. Compare figures 2 and 4 of Rau and figures 3 and 4 of Huber: Rau, "Das Quagga;" Huber, "Das Münchner Quagga."
75. Van Rensburg, "City Taxidermist."
76. National Archives and Records Service of South Africa, SAB Photo 17167.
77. Rau, "Revised List."
78. Edwards, *Gleanings of Natural History*, 29. Rau, "Revised List." Barnaby observed that the stripes on the body of the Milan quagga, which were evident in an old photograph, had become less distinct: Barnaby, *Quaggas*, 65.
79. Lydekker, "Note on the Quagga," 430–31. The reference to "York's picture" is the London mare shown in Figure 5.1.
80. This mare provided the opportunity to check Cornwallis Harris' claim that quaggas had four nipples: as noted previously, there were only two.
81. Dr. Carsten Renker, personal communication, May 2, 2018.
82. The farm was in Nelspoort in Beaufort West of the Cape Colony (now the Western Cape Province). Skead, *Historical Mammal Incidence*, 1:341.
83. Frost, "The Oven Bird."
84. Darwin and Stauffer, *Charles Darwin's Natural Selection*, 95, 166.
85. Gray, "Family Equidae."
86. Selous and Roosevelt, *African Nature Notes*, 131.
87. Selous, "Burchell's Zebra," 79.
88. Pocock, "Coloration;" Rau, "Revised List;" Larison et al., "How the Zebra." This cline in coat coloration across the plains zebra's range was described in Chapter 3.
89. Mayr, "What Is a Species?" The "Biological Species Concept" is frequently used but there exist other ways of characterizing species which are beyond the scope of this account.
90. Darwin had pondered whether hybridization might occur between zebras in nature, "We turn the Zebra into the Quagga, let them be wild in the same country with their own instinct (even though ~~fertile~~ hybrids produced when compelled to breed) ..." The crossed-out "~~fertile~~" in Darwin's notes reflects his ambivalence about the outcome of hybridization between quaggas and zebras, although it is unclear whether he was referring to plains zebras or mountain zebras. Darwin, "Notebook C," 145. Sterile offspring might argue that quaggas and zebras were separate species, for example, mules produced by crossing a male donkey with a horse mare are sterile; however, waterfowl of different species mate to give fertile hybrids: Ottenburghs et al., "Hybridization in Geese: A Review." And plains zebras interbreed with Grévy's zebras to produce fertile offspring: Cordingley et al., "Is the Endangered Grevy's Zebra Threatened by Hybridization?" Plains zebras and Grévy's zebras are thought to be valid species as they look quite different and so production of fertile hybrids surprised many biologists.

91. Lydekker, "Note on the Quagga," 428.
92. Pocock, "Cape Colony Quaggas," 317; Pocock, "Preorbital Pit," 518.
93. Lundholm, "Skull of the True Quagga." Cabrera also judged quaggas and plains zebras to be different species based on, "very noteworthy divergences" in the structure of their skulls: Cabrera, "Subspecific and Individual Variation," 91. Cabrera referenced two other biologists who had also reported differences in their skulls.
94. Rau, "Revised List."
95. Klein and Cruz-Uribe, "Craniometry of the Genus Equus."
96. Eisenmann and Brink, "Koffiefontein Quaggas;" Groves and Bell, "New Investigations."
97. Thackeray, "Zebras from Wonderwerk;" Thackeray, "Morphometric, Palaeoecological."
98. Rau, *Rough Road*, 2.
99. Lowenstein and Ryder, "Immunological Systematics."
100. Higuchi et al., "DNA Sequences."
101. Higuchi et al., "Mitochondrial DNA." An online account that provides details of this work can be accessed at Anon, "Phylogenetic Analysis."
102. Higuchi et al., "DNA Sequences;" Higuchi et al., "Mitochondrial DNA;" Heywood, "Micro-Politics of Macromolecules."
103. Klein and Cruz-Uribe, "Craniometry of the Genus Equus." The mitochondrial genome was later shown to be 16,366 nucleotides long in quaggas and so the 229 nucleotide long sequences were, indeed, just a small fraction of the whole genome: Vilstrup et al., "Mitochondrial Phylogenomics."
104. Leonard et al., "Loss of Stripes." Higuchi and his team had extracted DNA from muscle and connective tissue, but this later investigation sequenced DNA obtained from hides, a bone, and a tooth.
105. Leonard et al., "Loss of Stripes."
106. Vilstrup et al., "Mitochondrial Phylogenomics;" Jónsson et al., "Speciation with Gene Flow."
107. Jónsson et al., "Speciation with Gene Flow." The date of about 1.6 million years ago for the divergence of zebra species was also given by Steiner and Ryder, "Molecular Phylogeny." They suggested that mountain zebras split off at this time, followed by Grévy's zebras and plains zebras around 1.2 million years ago.
108. Jónsson et al., "Speciation with Gene Flow."
109. Pedersen et al., "Southern African Origin."
110. Hennessy, "Saving Species;" Fennessy et al., "Multi-Locus Analyses."
111. Pedersen et al., "Southern African Origin."
112. Poinar, "Ancient DNA."
113. Hofreiter, "Palaeogenomics."
114. Crichton, *Jurassic Park: A Novel.*
115. Groves and Bell, "New Investigations."
116. Ross, *Concise History*, 198–202.
117. Hendricks, "Remaking |Xam Narratives."
118. Anon, "Symbols." Reproduced with permission of the Western Cape Provincial Parliament.
119. Clasquin, "Quagga Kultuur," 1.
120. Anon, "Symbols."

Chapter 8

1. Harris, "Extinct Kin of Zebras." Rau expressed the same sentiment in Bisseker, "Genetic Science."
2. Other terms used for "rebreeding" include: "breed back," "re-breed," "reconstitute," "re-animate," "retrieve," "revive," "recreate," and "restore." The Quagga Project frequently uses the terms restore and restoration.
3. Tegetmeier et al., *Letters to Mr Tegetmeier*, 26.
4. Heck, "Breeding-Back." Lutz Heck is featured in the 2017 movie based in the Warsaw Zoo: Caro, "Zookeeper's Wife" (Focus Features, 2017).
5. De Bruxelles, "Shaggy Cow Story;" Kolbert, "Recall of the Wild."
6. "Heck cattle" and "Heck horses" occur at several locations, including the Hellabrunn Zoo in Munich, where Heinz Heck was director. Rebreeding is described in Anon, "Breeding-Back of the Tarpan;" Heck, *Animals, My Adventure*.
7. Heck, "Breeding-Back;" Thamm, "Rebirth of the Quagga."
8. Heck, *Animal Safari*, 85.
9. Rau, "Revised List;" Rau, "Additions to the Revised List."
10. Rau, *Rough Road*, 2. Rau's nine-page-long account written in 1999 is the source of much early information about the Quagga Project.
11. Rau, *Rough Road*, 2.
12. Rau, *Rough Road*, 4.
13. Rau, *Rough Road*, 4.
14. Thamm, "Rebirth of the Quagga," 209; Lundholm, "Rebirth of the Quagga?"
15. Higuchi et al., "Mitochondrial DNA."
16. Rau, *Rough Road*, 5.
17. Lowenstein and Ryder, "Immunological Systematics."
18. There are differing accounts of the membership of this committee: Anon, "Project Coordination;" Rau, *Rough Road*. Both sources include these members: Dr. M. A. Cluver, Professor E. H. Harley, Dr. P. A. Hulley and Mr. R. E. Rau. Rau additionally listed: Professor G. N. Louw, Dr. G. R. McLachlan, Mr. W. Morsbach, Dr. J. F. Warning, and Dr. J. A. van Zyl. However, under "Project Coordination," the Quagga Project website additionally lists: Professor C. W. Cruywagen, Mr. M. Gregor, and Professor H. J. Heydenrych.
19. March Turnbull, interview with the author, April 19, 2018.
20. Rau, *Rough Road*, 9.
21. Rau, "Quagga Experimental Breeding;" Friedel, "The Story of 'Howie'."
22. Rau, "Quagga Experimental Breeding;"
23. Barnaby, *Quaggas*.
24. Barnaby, "The Karoo Receives Plains Zebras."
25. Friedel, "The Story of 'Howie'."
26. Ferris, "Parks Body Joins Quest." The locations of these national parks are shown in Figure 0.2."
27. Anon, "Coordinator Reports," March 2017.
28. Anon, "Coordinator Reports," March 2018. Carol Freeman expressed concerns about the stresses experienced by zebras when pursued, tranquilized, and transported for relocation: Freeman, "Ending Extinction."
29. Anon, "Coordinator Reports," February 2015.

30. Anon, "Coordinator Reports," February 2019.
31. Darwin, *Variation of Animals and Plants*, vol. 2, 193.
32. Rau, "Quagga Project Management/Action Plan." The only version of this plan that I have been able to locate is described as "Draft" and is dated January 2005. The objective of breeding animals with "reddish" muzzles is curious as the muzzles of quaggas are described as black in some historic accounts, e.g., Harris, *Portraits of the Game*, 12; Smith, *Natural History*. The goal of "Body stripes not extending to the ventral midline" was to select animals with the white ventral surfaces characteristic of quaggas. Among the objectives of the plan was to have a population of "500 quagga-like zebras by 2020," but this number has not been achieved. The plan stated that there were more than fifty Quagga Project zebras located in national parks in 2004 and envisioned selection among these animals so that plains zebras with less quagga-like coat coloration would be removed from the parks; it is unclear whether this process has been carried out.
33. Mendel, "Pflanzen-Hybriden." The results of Mendel's work are described in modern terms as the words "genes" and "alleles" were introduced after his time.
34. Anon, "Objectives."
35. The size of the founder population was described as eighteen animals in Harley and Wooding, "Quagga Project FAQ," and nineteen animals in Harley et al., "Restoration of the Quagga."
36. A 2013 study found that inbreeding in Quagga Project zebras was "generally low" and that the coat coloration achieved by the fourth generation had been attained without affecting the genetic integrity of the animals: Hrabar and Kerley, *Selective Breeding*, 2.
37. Anon, "Coordinator Reports," February 2019.
38. Anon, "What Is a Quagga?"
39. Anon, "Project."
40. Anon, "Coordinator Reports," November 2006. The average price of less than R6,000 per animal would have been similar to the price of a regular plains zebra and indicates that these animals were being eliminated because they were not good breeding stock. Craig Lardner commented on the sale of such animals, "We sold off some animals without potential and gave the project a financial boost:" Ferreira, "Quagga Rebreeding," 53.
41. Anon, "Back from the Dead?" Anon, "Coordinator Reports," March 2017.
42. Anon, "Coordinator Reports," February 2015.
43. Bontebok Ridge Reserve is located close to Elandsberg and comprises 40 hectares of vineyard and approximately 400 hectares of private reserve where, in addition to the zebras, there are wildebeests, springboks, elands, and boneteboks. Turner Corporation, "Bontebok Ridge Reserve." The Bontebok Ridge Reserve is dedicated to preserving animals and the native renosterveld vegetation that sustains them. In April 2018, I had planned to interview the owners of the Bontebok Ridge Reserve, Tom and Katja Turner, and visit their reserve, but rain – welcome at that drought-stricken time – prevented a visit.
44. Harley and Wooding, "Quagga Project FAQ."
45. Harley and Wooding, "Quagga Project FAQ."
46. Darwin wrote, "The Quagga, though strongly banded in the front part of the body is without stripes on the legs; but one individual which Lord Derby kept alive had a few distinct zebra-like transverse bars on the hocks." Darwin and Stauffer, *Charles Darwin's Natural Selection*, 328–29.

47. The quotation is from Anon, "What Is a Quagga?" The evaluation was from Parsons, Aldous-Mycock, and Perrin, "A Genetic Index." Turnbull also commented on the light brown color of Quagga Project zebras and advocated that a darker brown color of the upper body should be continuing goals of the Quagga Project: Turnbull, "Back from the Dead."

48. Eric Harley, quoted in Anon, "Coordinator Reports," August 2015, 9.

49. Boddaert, Sistens Quadrupedia, 160; von Linné and Gmelin, Systema Naturæ.

50. Edwards, Gleanings of Natural History, 29.

51. Somerville, Bradlow, and Bradlow, William Somerville's Narrative, 36; Harris, Portraits of the Game, 12.

52. Anon, "Coordinator Reports," August 2020.

53. Harley et al., "Restoration of the Quagga," 80. This statement contrasts with a less determined attitude towards developing brown coloration expressed on the Quagga Project website. In comparing images of a third generation Quagga Project zebra with two quagga museum specimens, the website states, "The background brown colour is not so well developed as in either museum example, but this is such a variable feature in the museum specimens (of which there are 24 in all), that this is only a secondary concern for the project" Anon, "What Is a Quagga?" As discussed in Chapter 7, the variation in brown coloration of museum specimens probably reflects different degrees of fading, and so it is incongruous to compare Rau quaggas with faded museum specimens as a justification for not giving development of brown coloration a high priority.

54. Bernard Wooding, interview with the author, April 19, 2018.

55. March Turnbull, personal communication, June 21, 2021.

56. Cary [Mungall], "History and Extinction of the Quagga." 44.

57. Dekkers, "Application of Genomics;" Meuwissen, Hayes, and Goddard, "Genomic Selection."

58. Parsons, Aldous-Mycock, and Perrin, "A Genetic Index," 110.

59. Pedersen et al., "Southern African Origin." Anon, "Coordinator Reports," March 2018.

60. Anon, "Coordinator Reports," June 2008.

61. Anon, "Coordinator Reports," August 2013.

62. Anon, "Coordinator Reports," August 2020. March Turnbull, personal communication, December 8, 2021.

63. Wildswinkel Boland Veiling.

64. Child, "Quaggas Come Back to Life." Probably sellers and buyers knew that the mares were pregnant.

65. March Turnbull, interview with the author, April 19, 2018.

66. Wildswinkel Boland Veiling; Anon, "Coordinator Reports," March 2017.

67. Wildswinkel Boland Veiling.

68. Anon, "Back from the Dead?"

69. Graff, "Shifting Sands."

70. Dr. Seuss, If I Ran the Zoo.

71. March Turnbull, interview with the author, April 19, 2018.

72. Anon, "Coordinator Reports," June 2021. It was reported that thirty-six foals had been born in the preceeding twenty-four months. The numbers of Core Herd animals at each location in June 2021 are represented by the first figure in the

parentheses; the possible future numbers of zebras proposed by the Quagga Project Association are the second figure in the parentheses: Elandsberg (30, 30), the Nuwejaars Wetlands Special Management Area (25, 50), Pampoenvlei (25, 25), and Vlakkenheuwel (26, 25). In June 2021, the 112 animals of the Quagga Core herd also included six animals at other locations: March Turnbull, personal communication, June 20, 2021. In July 2020, there were approximately sixty animals outside the Quagga Core Herd that belonged to a breeder: March Turnbull, personal communication, August 6, 2020.

73. Barnard, "African Horse Sickness."
74. March Turnbull and Bernard Wooding, interview with the author, April 19, 2018.
75. Bartholomeus Klip features in the Netflix drama *The Crown* as the Kenyan guesthouse where Princess Elizabeth learned of the death of her father, King George VI.
76. March Turnbull, interview with the author, April 19, 2018. Rau himself envisioned using artificial insemination followed by embryo transplants into surrogate equine mothers: Rau, "Quagga Experimental Breeding," 436.
77. Eric Harley's Ancestor Contribution Program shows that Alex had 292 descendants as of September 2017. Other founders with many descendants include Lulu, with 261, Melanie, with 252 and Allan, with 239.
78. March Turnbull, interview with the author, April 19, 2018.
79. Genetic evidence indicates gene exchange between plains zebras and mountain zebras. Pedersen et al., "Southern African Origin."
80. Anon, "Coordinator Reports," February 2015. Angela Gaylard was quoted in an article reproduced on page 22 of the February 2015 Coordinator's Report.
81. March Turnbull, interview with the author, April 19, 2018.
My enquiries to officials at SANParks regarding their attitudes to Quagga Project zebras have not been answered.
82. Although SANParks is not accepting further Quagga Project zebras, the websites for the Karoo National Park and Addo Elephant National Park both refer to plains zebras with reduced stripes, and so it appears that Quagga Project zebras and their offspring remain in these national parks.
83. March Turnbull, personal communication, August 6, 2020.
84. Anon, "Coordinator Reports," August, 2020.
85. There was an advertisement with similar – but not identical – wording in *Sagittarius: Magazine of the South African Museum* 4 (2) (1989), 24.
86. Anon, "Sponsors;" Rau, *Rough Road*.
87. Anon, "Coordinator Reports," March 2018.
88. Anon, "Coordinator Reports," February 2014.
89. Anon, "Sponsors."
90. Yeld, "Re-Breeding of Quagga." This sum is the only budget information that I have encountered but see below for an estimate of the current budget.
91. Anon, "Coordinator Reports," March 2018. Expenditure of the R1,740,000 realized by the 2017 auction over slightly more than four years points to a current annual budget in the range of R400,000 to R500,000 (equivalent to US$27,880 to US$34,860).
92. Anon, "Hunters Back Bid to Rebreed the Quagga." The Quagga Project website lists an additional five hunting organizations that have donated.

93. Streak, "Quagga Project Forced;" Rau also observed that CHASA was involved in conservation projects.
94. Wedderwill, "Quagga Project"
95. Rau, *Rough Road*, 2.
96. Seddon, Moehrenschlager, and Akcakaya, "IUCN Guiding Principles on Creating Proxies."
97. Stokstad et al., "Bringing Back the Aurochs."
98. Miller et al., "Identification."
99. Phillips, Henry, and Kelly, "Restoration of the Red Wolf," 273.
100. Phillips, Henry, and Kelly, "Restoration of the Red Wolf."
101. Manganiello, "From a Howling Wilderness."
102. Gurdon, "Transplanted Nuclei."
103. Wilmut et al., "Viable offspring."
104. Johnson, "Clone of the Endangered Przewalski's Horse."
105. Anon, "Frozen Zoo."
106. Novak, "De-Extinction." Ronald Sandler has described the prospect of recreating long extinct species as "deep de-extinction" and has provided a thoughtful discussion of its ethics: Sandler, "Ethics of Reviving." Candidates for deep de-extinction include mammoths (*Mammuthus primigenius*), passenger pigeons (*Ectopistes migratorius*), and thylacines (*Thylacinus cynocephalus*). See also Seddon, Moehrenschlager, and Akcakaya, "IUCN Guiding Principles on Creating Proxies."
107. Novak, "De-Extinction."
108. Harris, "Extinct Kin of Zebras."
109. *Wildswinkel Boland Veiling*. This video conveys the excitement generated at the auction of Quagga Project zebras.
110. Jordan, "Karoo Welcomes Quagga Back from the Dead;" McNeil, "Brave Quest." Unfortunately, both headlines imply that Quagga Project zebras *are* quaggas.
111. Anon, "Project."
112. Anon, *Rebuilding a Species*.
113. De Vos, "Stripes Faded."
114. Macnab, "Quagga's World." Biologists are increasingly aware of the importance of microbiota, for example, bacteria in the intestines of animals. A variety of small organisms also live on the surfaces of animal bodies in such specialized environments as hair follicles. Extinction of the host may lead to loss of another organism unique to that species or subspecies, and this state of affairs may have been the case for quaggas.
115. Rau, "Quagga Project Management/Action Plan," 2, 1.
116. Darwin, *Variation of Animals and Plants*.

Chapter 9

1. Anon, "Back from the Dead?"
2. Pringle, *Narrative of a Residence*, 149.
3. Rau, "Quagga Experimental Breeding," 436.
4. The biologist P. J. H. van Bree made a similar comment with regard to Napoleon Bonaparte. Max, "Can You Revive an Extinct Animal?"

5. Boroughs, "Stripes and Shadows," 49.
6. Anon, "Project."
7. Anon, "Project."
8. Gray, "Family Equidae;" Boddaert, *Sistens Quadrupedia*, 160.
9. Skinner, in his 1996 paper, cited both Higuchi et al., "DNA Sequences" and Lowenstein, "Half-Striped Quaggas." Nevertheless, he judged that morphological evidence indicated that quaggas were a distinct species and were more closely related to mountain zebras than to plains zebras: Skinner, "Further Light."
10. Klein and Cruz-Uribe, "Craniometry of the Genus Equus;" Lundholm, "Skull of the True Quagga;" Thackeray, "Morphometric, Palaeoecological." The skull (TM 10161 in the Ditsong Museum) listed on the Quagga Project website as one of the quagga artifacts in South Africa was identified solely on the basis of morphological differences that exist between the skulls of quaggas and other plains zebras, see Chapter 7.
11. Klein and Cruz-Uribe, "Craniometry of the Genus Equus," 86.
12. Heywood, "Sexual Dimorphism."
13. Pedersen et al., "Southern African Origin," 491. The "[was]" in the quote was approved by Casper-Emil Pedersen, personal communication, December 13, 2018.
14. Anon, "Coordinator Reports," March 2018, 10.
15. Jónsson et al., "Speciation with Gene Flow." Additional details of this genomic research are described in Chapter 7.
16. Babies produce the enzyme lactase that enables them to break down lactose and digest their mothers' milk, but in some human populations, lactose-tolerance ceases in childhood and milk becomes indigestible: Hassan et al., "Genetic Diversity."
17. Hedrick, "Population Genetics."
18. Eric Harley, quoted in Anon, "Coordinator Reports," March 2016, 10.
19. Anon, "View Quaggas."
20. Anon, "Objectives."
21. Anon, "Coordinator Reports," August 2013, 7. Another instance where Quagga Project zebras are identified by a trinomial name occurs in a scanned copy of an article on the Quagga Project website, "Today, the project has 100 animals in its breeding programme, with 14 of them classified as Equus quagga quagga." Ferreira, "Quagga Rebreeding," 52.
22. Child, "Quaggas Come Back to Life."
23. Anon, "What Is a Quagga?"
24. Anon, "Karoo National Park;" Anon, "Addo Elephant National Park." The "pale rumps" in this description do not do justice to the chestnut-colored hindquarters of quaggas.
25. Pettorelli, Durant, and du Toit, *Rewilding*; Navarro and Pereira, "Rewilding;" Kolbert, "Recall of the Wild."
26. Pettorelli, Durant, and du Toit, *Rewilding*.
27. Alston et al., "Reciprocity in Restoration Ecology;" Hayward et al., "Top-Down Control," 335.
28. Beschta and Ripple, "Riparian Vegetation Recovery."
29. Seddon and Armstrong, "Translocation in Rewilding;" Gaywood, "Reintroducing the Eurasian Beaver."
30. Kivinen, Nummi, and Kumpula, "Beaver-Induced."

31. Johns, "History of Rewilding," 27; Linnell and Jackson, "Bringing Back Large Carnivores."
32. Rau, "Additions to the Revised List," 44.
33. Brakes et al., "Animal Cultures Matter for Conservation;" Whiten, "Cultural Evolution in Animals."
34. Freeman also questions whether there will be a large enough population of Quagga Project zebras to fill the environmental role once occupied by quaggas or merely a smaller number of animals as a lure for tourists: Freeman, "Ending Extinction."
35. Greig, "Lesson of the Quagga."
36. Boroughs, "Stripes and Shadows," 49.
37. Masubelele et al., "Vegetation Change."
38. March Turnbull and Bernard Wooding, interview with the author, April 19, 2018.
39. Professor Cluver, director of the South African Museum and a founding member of the Quagga Project, was quoted in Bisseker, "Genetic Science."
40. Rau, "Quagga Experimental Breeding," 436. The auctioneer at the March 17, 2017 auction of three Quagga Project zebras also made the claim, "This is true conservation . . ." *Wildswinkel Boland Veiling.*
41. Rau, *Rough Road*, 2, 5.
42. Boroughs, "Stripes and Shadows," 44.
43. Endangered Wildlife Trust, "Colour variants."
44. Dr. Andrew Taylor, personal communication, January 5, 2021.
45. Stears, Shrader, and Castley, "Equus quagga ".
46. Dr. Keenan Stears, personal communication, July 17, 2020. Striping abnormalities in inbred plains zebras are described in Larison et al., "Population Structure."
47. Endangered Wildlife Trust, "2016 Red List."
48. Anon, "Threatened Species Programme."
49. An example of making choices in conservation was that the Hawaiian goose (*Branta sandvicensis*) had a much higher priority than another endangered Hawaiian bird: Wilson, "Making the Nēnē Matter."
50. Bennett et al., "Spending Limited Resources."
51. Freeman, "Ending Extinction;" Nicole Itano, "S. Africa's Quagga Saga," 7; Page and Hancock, "Zebra Cousin Went Extinct."
52. Lorimer, *Wildlife.*
53. Beilfuss et al., "Status and Distribution," 47. Selous' zebras lack brown shadow stripes and have conspicuous black and white stripes on their legs and bodies.
54. Dutton and Dutton, "Tragedy in the Making;" Stalmans et al., "War-Induced Collapse."
55. Anon, "Selous Zebra."
56. March Turnbull, interview with the author, April 19, 2018.
57. Botanists have identified 850 species of flowering plants on Elandsberg Reserve; four are endemic to the Reserve and approximately 100 are endangered. SANBI, the South African Biodiversity Institute of the Kirstenbosch National Botanical Garden in Cape Town, has collected seeds at Elandsberg and has grown these rare plants at other locations. The geometric tortoise faces diverse challenges including fire and predation by crows, which eat eggs and young. Volunteers rescued these tortoises after a fynbos fire some years ago; now they are bred at Elandsberg Reserve, and their young are raised in an enclosure protected by netting against crows until they are old enough to be released safely to the veld.

58. Rau's advocacy for geometric tortoises led to the formation of the Eenzaamheid Provincial Nature Reserve: Boroughs, "Stripes and Shadows."
59. Harley et al., "Whole Mitochondrial Genome."
60. Star and Griesemer, "Institutional Ecology."
61. Anon, "Grevy's Zebra Trust." The Grevy's Zebra Trust does not place an accent on the name.
62. Antonius, "Geographical Distribution of Recent Equidæ," 564.
63. The quotation is from Donald Waller, recorded in Popkin, "Can Genetic Engineering Bring Back the American Chestnut?" The American chestnut, *Castanea dentata*, was a common and economically important tree in the northeastern United States until ravaged by an imported fungal blight early in the twentieth century. Thanks to the American Chestnut Foundation and its supporters, there are now blight-resistant trees created by both conventional breeding techniques and by recombinant DNA technology.

Appendix 1

1. Rock engravings n.d., Fock and McGregor, "Two Rock Engravings," Figures 1 and 2; M. Wilman, *Rock-Engravings of Griqualand West & Bechuanaland South Africa*, Plate 23; de Prada Samper, personal communication; Edwards, *Gleanings of Natural History*, Plate 223; Gordon, *The Gordon African Collection* (RP-T-1914-17-190); Schouman, *Een Quagga (Equus quagga)*, original is in Teylers Museum, Haarlem, Netherlands; Maréchal, "Quagga," available from https://commons.wikimedia.org/wiki/File: Quagga.jpg; and Dorst, "Notice Sur Les Spécimens;" Daniell, *Collection of Plates Illustrative of African Scenery and Animals*, Plate 15; Agasse (1817) in Le Fanu, *Catalogue*; Kehrer in Schlawe, *Steppenzebras Von Sudafrika*, Figure 23; Agasse (1821) in Le Fanu, *Catalogue*; Sowerby in Gall, "Illustrations of the Quagga," Figure 8; Harris, *Portraits*, Plate 2; Smith, *Natural History*, Plate 24; Wagner in Huber, "Das Münchner Quagga," 157; Hawkins in Gray et al., *Gleanings from Knowsley Hall*, Plate 54; Weir, "Quagga"; Zimmermann in Schlawe, *Steppenzebras Von Sudafrika*, Figure 12; Haes (1863, 1864) in Huber, "Documentation of the Five Known Photographs," Figures 4 and 5; Fritsch, *Drei Jahre in Süd-Afrika*, Plate 25; York in Huber, "Documentation of the Five Known Photographs," Figures 2 and 3.

Bibliography

Adhikari, M. *Anatomy of a South African Genocide: The Extermination of the Cape San Peoples*. Athens, OH: Ohio University Press, 2011.

Allen, D. S. E. "The Quagga." *The Times*. November 24, 1951, 7.

Alston, J. M., B. M. Maitland, B. T. Brito, S. Esmaeili, A. T. Ford, B. Hays, . . . J. R. Goheen. "Reciprocity in Restoration Ecology: When Might Large Carnivore Reintroduction Restore Ecosystems?" *BIOC Biological Conservation* 234 (2019): 82–89.

Anon. "Acute Indigestion, Gorged Stomach, or Stomach Staggers of the Horse." *The Agricultural Journal of the Cape of Good Hope* 15, no. 9 (1899): 590–91.

"Back from the Dead? First Quagga Project Sales Reach Record Prices at Auction." The Quagga Project, 2017. https://quaggaproject.org/first-quagga-project-sales-.

"Bontebok Damaliscus pygargus." *San Diego Zoo Wildlife Alliance*, 2020. https://animals.sandiegozoo.org/animals/bontebok, last accessed December 8, 2020.

"Breeding-Back of the Tarpan." *Nature* 171, no. 4362 (1953): 1008.

"Coordinator Reports." The Quagga Project, n.d. https://quaggaproject.org/project-coordination/.

"Exit Quagga." *Fur Trade Review* 15, no. 2 (1887): 100.

"For Wild Zebra ~ It's All About the Spots." *Simply Marvelous Horse World*, 2012. https://simplymarvelous.wordpress.com/tag/spotted-zebra/, last accessed August 10, 2017.

"Frozen Zoo." n.d. https://sciences.sandiegozoo.org/resources/frozen.

"Grevy's Zebra Trust." n.d. http://www.grevyszebratrust.org.

"Harvard-MIT Mathematics Tournament." 2003. www.hmmt.co/static/archive/february/solutions/2003/sguts03.pdf, last accessed October 22, 2016.

"Hunters Back Bid to Rebreed the Quagga." *The Argus* (South Africa). September 17, 1987.

"Jan Christoffel Greyling Kemp (1872–1946)." *WikiTree*, n.d. www.wikitree.com/wiki/Kemp-8046, last accessed November 17, 2021.

"Karoo National Park." *South African National Parks*, n.d. www.sanparks.org/parks/karoo/conservation/ff/mammals.php, last accessed December 5, 2021.

"Laws Too Late to Save Quagga." *The Argus*. March 28, 1987.

"Objectives." The Quagga Project, n.d. https://quaggaproject.org/objectives/.

"The Project." The Quagga Project, n.d. https://quaggaproject.org/the-project/.

"The Quagga." *The Victorian Naturalist* 18 (1901): 16.

"Quagga Not Extinct. Herds in South-West Africa. Convincing Story from Namib." *The Star* (Johannesburg). June 18, 1932.

"The Quagga Project." 1987–2020. https://quaggaproject.org/.

"The Quagga Project." Wedderwill South Africa, updated January 17, 2011. http://wedderwill.co.za/category/blog/.

"Rebuilding a Species." *Voice of America News*, December 1, 2008.

"Selous Zebra." Blue Forest Safaris, n.d. www.wild-about-you.com/GameSelousZebra.htm.

"Sponsors." The Quagga Project, n.d. https://quaggaproject.org/sponsors/.

"Symbols." Western Cape Provincial Parliament, 2013. www.wcpp.gov.za/symbols, last accessed December 7, 2020.

"Threatened Species Programme." Red List of South African Plants, 2020. http://redlist.sanbi.org/.

"To Search for Quagga in Remote Hills. Expedition to Leave Soon for S.W. Africa." *The Cape Argus*. March 8, 1952.

"View Quaggas." The Quagga Project, n.d. https://quaggaproject.org/view-quaggas/.

"What Is a Quagga?" The Quagga Project, n.d. https://quaggaproject.org/what-is-a-quagga/.

"What Is a Quagga? Phylogenetic Analysis of the Quagga." Allan Wilson Centre, 2015. www.allanwilsoncentre.ac.nz/massey/learning/departments/centres-research/allan-wilson-centre/our-research/resources/recreate-the-research/what-is-a-quagga/phylogenetic-analysis-of-the-quagga.cfm.

"The Quagga." *The Times*. November 24, 1951, 7.

Antonius, O. "On the Geographical Distribution, in Former Times and Today, of the Recent Equidæ." *Proceedings of the Zoological Society of London* B107, no. 4 (1938): 557–64.

Baard, E. H. W. "Distribution and Status of the Geometric Tortoise Psammobates geometricus in South Africa." *Biological Conservation* 63, no. 3 (1993): 235–39.

Baines, T. "The Greatest Hunt in History near Bloemfontein 1860." 1861. https://commons.wikimedia.org/wiki/File:The_greatest_hunt_in_history_near_Bloemfontein_1860.jpg, last accessed October 16, 2021.

Journal of Residence in Africa, 1842–1853. 2 vols. Cape Town: Van Riebeeck Society, 1961–64.

Baker, S. W. "William Cotton Oswell: A Biographical Sketch." In *Big Game Shooting.* Edited by C. Phillipps-Wolley, 28–29. London: Longmans, Green & Co., 1894.

Bales, S. L. *Ghost Birds: Jim Tanner and the Quest for the Ivory-Billed Woodpecker, 1935–1941.* Knoxville: University of Tennessee Press, 2010.

Bank, A. "Anthropology and Portrait Photography: Gustav Fritsch's 'Natives of South Africa,' 1863–1872." *Kronos* (Bellville, South Africa) 27 (2001): 43–76.

Bushmen in a Victorian World: The Remarkable Story of the Bleek-Lloyd Collection of Bushman Folklore. Cape Town: Double Storey, 2006.

Bard, J. B. L. "A Model for Generating Aspects of Zebra and Other Mammalian Coat Patterns." *Journal of Theoretical Biology* 93, no. 2 (1981): 363–85.

"A Unity Underlying the Different Zebra Striping Patterns." *Journal of Zoology* 183, no. 4 (1977): 527–39.

Barnaby, D. "The Karoo Receives Plains Zebras from the Quagga Project." *International Zoo News*, no. 291 (1999): 94–98.

Quaggas and Other Zebras, Plymouth: Basset, 1996.

Barnaby, D., J. Adams, and Bartlett Society. *Quagga Quotations: A Quagga Bibliography*. Southampton: Bartlett Society, 2001.

Barnard, A. *Bushmen: Kalahari Hunter-Gatherers and Their Descendants*. Cambridge: Cambridge University Press, 2019.

Barnard, B. J. "Epidemiology of African Horse Sickness and the Role of the Zebra in South Africa." *Archives of Virology. Supplementum* 14 (1998): 13–19.

Barnett, R., N. Yamaguchi, I. Barnes, and A. Cooper. "Lost Populations and Preserving Genetic Diversity in the Lion Panthera leo: Implications for Its Ex Situ Conservation." *Conservation Genetics* 7, no. 4 (2006): 507–14.

Barrow, J. *An Account of Travels into the Interior of Southern Africa, in Years 1797 and 1798*. 2 vols. London: T. Cadell, 1801–04.

Beddard, F. E. *Natural History in Zoological Gardens*. London: A. Constable & Co., 1905.

Beilfuss, R. D., C. M. Bento, M. Haldane, and M. Ribaue. *Status and Distribution of Large Herbivores in the Marromeu Complex of the Zambezi Delta, Mozambique*. Maputo, Mozambique: World Wide Fund for Nature, 2010.

Beinart, W. "The Night of the Jackal: Sheep, Pastures and Predators in the Cape." *Past and Present*, no. 158 (1998): 172–206.

Bennett, J. R., R. F. Maloney, T. E. Steeves, J. Brazill-Boast, H. P. Possingham, and P. J. Seddon. "Spending Limited Resources on De-Extinction Could Lead to Net Biodiversity Loss." *Nature Ecology & Evolution* 1, no. 4 (2017): 1–4.

Berger, J. *Ways of Seeing*. London: British Broadcasting Corporation, 2008.

Berridge, W. S., and W. P. Westell. *The Book of the Zoo*. London: J. M. Dent & Sons, 1911.

Beschta, R. L., and W. J. Ripple. "Riparian Vegetation Recovery in Yellowstone: The First Two Decades after Wolf Reintroduction." *BIOC Biological Conservation* 198 (2016): 93–103.

Bisseker, C. "Genetic Science Helps Right the Wrongs of the Past." *Financial Mail* 158, no. 4 (2000): 41.

Bleek, D. F. *A Bushman Dictionary*. New Haven, CN: American Oriental Society, 1956.

Bleek, W. H. I., and L. Lloyd. "The Digital Bleek and Lloyd." http://lloydbleekcollection.cs.uct.ac.za.

Bleek, W. H. I., L. Lloyd, and G. McCall Theal. *Specimens of Bushman Folklore*. London: G. Allen & Co., 1911.

Boddaert, P. *Sistens Quadrupedia*. Elenchus Animalium, vol. 1, Roterodami: C. R. Hake, 1785.

Boroughs, D. "Stripes and Shadows." *Timbila: Spirit of Africa* 2, no. 1 (2000): 42–49.

Boshoff, A. F., and L. J. Kerley, "Potential Distributions of the Medium- to Large-Sized Mammals in the Cape Floristic Region, Based on Historical Accounts and Habitat Requirements." *African Zoology* 36, no. 2 (2001): 245–73.

Boshoff, A., M. Landman, and G. Kerley, "Filling the Gaps on the Maps: Historical Distribution Patterns of Some Larger Mammals in Part of Southern Africa." *Transactions of the Royal Society of South Africa* 71, no. 1 (2016): 23–87.

Boyle, C. L. "The Quagga." *The Times*. August 9, 1951, 5.

Bradfield, J. *All That Glitters*. Scotts Valley, CA: CreateSpace Publishing, 2010.

Brakes, P., S. R. X. Dall, L. M. Aplin, S. Bearhop, E. L. Carroll, P. Ciucci, ... C. Rutz. "Animal Cultures Matter for Conservation." *Science* 363, no. 6431 (2019): 1032–34.

Brubaker, A. S., and R. G. Coss. "Evolutionary Constraints on Equid Domestication: Comparison of Flight Initiation Distances of Wild Horses (Equus caballus ferus) and Plains Zebras (Equus quagga)." *Journal of Comparative Psychology* 129, no. 4 (2015): 366–76.

Bryden, H. A. "The Blesbok." In *Great and Small Game of Africa*. Edited by H. A. Bryden, 183–90. London: R. Ward, 1899.

Kloof and Karroo: Sport, Legend and Natural History in Cape Colony. London: Longmans, Green & Co., 1889.

"The Quagga (Equus quagga)." In *Great and Small Game of Africa*. Edited by H. A. Bryden, 72–79. London: R. Ward, 1899.

Buckley, T. E. "On the Past and Present Geographical Distribution of the Large Mammals of South Africa." *Proceedings of the Zoological Society of London* 44, no. 1 (1876): 277–93.

Burchell, W. J. *Travels in the Interior of Southern Africa*. 2 vols. London: Printed for Longman, Hurst, Rees, Orme, and Brown, Paternoster-Row, 1822–24.

Burkhardt, R. W. "Closing the Door on Lord Morton's Mare: The Rise and Fall of Telegony." *Studies in History of Biology* 3 (1979): 1–21.

Burns, P. R. "Quagga Stamps" Cryptozoology and Philately, 2003. www.pibburns.com/cryptost/quagga.htm.

Büttiker-Otto, W. *Memories of a Scientist: The Carp Expedition to the Save River in Zimbabwe and Mozambique*. Basel: Basler Afrika Bibliographien, 2008.

Cabrera, A. "Subspecific and Individual Variation in the Burchell Zebras." *Journal of Mammalogy* 17, no. 2 (1936): 89–112.

Caccone, Gisella. "How a Zebra Lost Its Stripes: Rapid Evolution of the Quagga." *Yale News*, 2005. https://news.yale.edu/2005/09/26/how-zebra-lost-its-stripes-rapid-evolution-quagga.

Cain, J. W., N. Owen-Smith, and V. A. Macandza. "The Costs of Drinking: Comparative Water Dependency of Sable Antelope and Zebra." *Journal of Zoology* 286, no. 1 (2012): 58–67.

Cape of Good Hope. "Proclamation by His Excellency General the Right Honorable Lord Charles Somerset, 21 March 1822." In *Records of the Cape Colony*. Edited by G. M. Theal, 150–55. Printed for the Government of the Cape Colony, 1905.

Caro, N. *The Zookeeper's Wife*. Directed by Niki Caro. Focus Features, 2017.

Caro, T. *Zebra Stripes*. Chicago: The University of Chicago Press, 2016.

Caro, T., A. Izzo, R. C. Reiner, Jr., H. Walker, and T. Stankowich. "The Function of Zebra Stripes." *Nature Communications* 5 (2014): 3535.

Caro, T., and T. Stankowich. "Concordance on Zebra Stripes: A Comment on Larison *et al.*" *Royal Society Open Science* 2 (2015): 150323.

Carp, B. *I Chose Africa*. Cape Town: H. Timmins, 1961.

"A Quagga Chase in South-West Africa." *African Wild Life* 6, no. 2 (1952): 100–5.

Carpenter, K. "Halaelurus quagga (Alcock, 1899), Quagga Catshark." *Fishbase*, 2017. www.fishbase.se/summary/824, last accessed October 14, 2021.

Carruthers, J. "Changing Perspectives on Wildlife in Southern Africa, c.1840 to c.1914." *Society and Animals* 13, no. 3 (2005): 183–99.

National Park Science: A Century of Research in South Africa. Cambridge: Cambridge University Press, 2017.

Cary [Mungall], E. R. "History and Extinction of the Quagga: Barking Horse of the Karroo." BS Dissertation, University of Wisconsin, Madison, 1970.

Chadwick, P. "Saving Stripes." *Africa Geographic* 13, no. 2 (2005): 60–66.

Child, K. "Quaggas Come Back to Life." *The Times (South Africa).* May 15, 2017.

Clarke, J. *Overkill: The Race to Save Africa's Wildlife.* Cape Town: Struik Nature, 2017.

Clasquin, M. *Quagga Kultuur: Reflections on South African Popular Culture.* Wierda Park, Gauteng: Auroroa Press, 2003.

Cloudsley-Thompson, J. L. "Frederick Courteney Selous." *The Linnean* 16, no. 2 (2000): 24–31.

Cobb, A., and S. Cobb. "Do Zebra Stripes Influence Thermoregulation?" *Journal of Natural History* 53, no. 13–14 (2019): 863–79.

Copeland, S. R., H. C. Cawthra, E. C. Fisher, R. M. Cowling, P. J. le Roux, J. Hodgkins, and C. Marean. "Strontium Isotope Investigation of Ungulate Movement Patterns on the Pleistocene Paleo-Agulhas Plain of the Greater Cape Floristic Region, South Africa." *Quaternary Science Reviews* 141 (2016): 65–84.

Cordingley, J. E., S. R. Sundaresan, I. R. Fischhoff, B. Shapiro, J. Ruskey, and D. I. Rubenstein. "Is the Endangered Grevy's Zebra Threatened by Hybridization? Occurrence of Hybridization between Two Wild Zebra Species." *Animal Conservation* 12, no. 6 (2009): 505–13.

Cowie, H. *Exhibiting Animals in Nineteenth-Century Britain: Empathy, Education, Entertainment.* Basingstoke: Palgrave Macmillan, 2014.

Crais, C. C., and P. Scully. *Sara Baartman and the Hottentot Venus: A Ghost Story and a Biography.* Princeton, NJ: Princeton University Press, 2009.

Crean, A. J., A. M. Kopps, and R. Bonduriansky. "Revisiting Telegony: Offspring Inherit an Acquired Characteristic of Their Mother's Previous Mate." *Ecology Letters* 17, no. 12 (2014): 1545–52.

Crichton, M. *Jurassic Park: A Novel.* New York: Alfred Knopf, 1990.

Cucchi, T. A. Mohaseb, S. Peigné, K. Debue, L. Orlando, and M. Mashkour. "Detecting Taxonomic and Phylogenetic Signals in Equid Cheek Teeth: Towards New Palaeontological and Archaeological Proxies." *Royal Society Open Science* 4, no. 4 (2017): 160997.

de Cuvier, G. L. C. F. D. *The Animal Kingdom Arranged in Conformity with Its Organization, Vol. 3. The Class Mammalia.* London: Whittaker, 1827.

Daniell, S. *African Scenery and Animals, 1804–1805.* London: W. Daniell, 1820.

A Collection of Plates Illustrative of African Scenery and Animals. London: Samuel Daniell, 1804. Smithsonian Libraries SIL-SIL28–276-06. https://library.si.edu/image-gallery/106385.

Darwin, C. *The Descent of Man and Selection in Relation to Sex: Charles Darwin,* Vol. 1. London: John Murray, 1888.

"Notebook C." *The Complete Work of Charles Darwin Online*. Edited by J. Van Wyhe. http://darwin-online.org.uk/content/frameset?keywords=quagga&pageseq=129&itemID=CUL-DAR122.-&viewtype=text.

The Origin of Species. London: John Murray, 1859.

The Variation of Animals and Plants under Domestication, Vol. 2. London: Murray, 1868.

Darwin, C., and F. Darwin. *The Foundation of the Origin of Species, Two Essays Written in 1842 and 1844, by Charles Darwin edited by His Son Francis Darwin*. Cambridge: Cambridge University Press, 1909.

Darwin, C., and R. C. Stauffer. *Charles Darwin's Natural Selection: Being the Second Part of His Big Species Book Written from 1856 to 1858*. London: Cambridge University Press, 1975.

Davidson, Z., M. Dupuis-Desormeaux, A. Dheer, L. Pratt, E. Preston, S. Gilicho, . . . C. P. Doncaster. "Borrowing from Peter to Pay Paul: Managing Threatened Predators of Endangered and Declining Prey Species." *PeerJ*, no. 7 (2019): e7916.

Davis, J. "Eugen Sandow: A Body Worth Immortalising." Collections, Natural History Museum, n.d. www.nhm.ac.uk/discover/eugen-sandow-a-body-worth-immortalising.html.

De Bruxelles, S. "A Shaggy Cow Story: How a Nazi Experiment Brought Extinct Aurochs to Devon." *The Times*. April 22, 2009.

De Prada-Samper, J. M. "Roads, Ghosts and Rock Paintings in the Swartruggens, Western Cape Province." *The Digging Stick* 36, no. 1 (2019): 5–9.

"The Forgotten Killing Fields: 'San' Genocide and Louis Anthing's Mission to Bushmanland, 1862–1863." *Historia* 57, no. 1 (2012): 172–87.

De Prada-Samper, J. M., and J. de Villiers. *The Man Who Cursed the Wind and Other Stories from the Karoo*. Cape Town: African Sun Press, 2016.

De Vos, R. "Stripes Faded, Barking Silenced: Remembering Quagga." *Animal Studies Journal* 3, no. 1 (2014): 29–45.

Dean, W. R. J., and S. J. Milton. "Animal Foraging and Food." In *The Karoo: Ecological Patterns and Processes*. Edited by W. R. J. Dean and S. J. Milton, 164–78. Cambridge: Cambridge University Press, 1999.

Dekkers, J. C. M. "Application of Genomics Tools to Animal Breeding." *Current Genomics* 13, no. 3 (2012): 207–12.

Desmet, P. G., and R. M. Cowling. "The Climate of the Karoo: A Functional Approach." In *The Karoo: Ecological Patterns and Processes*. Edited by W. R. J. Dean and S. J. Milton, 3–16. Cambridge: Cambridge University Press, 1999.

Diamond, J. *Guns, Germs, and Steel: The Fates of Human Societies*. New York: Norton, 1997.

Dohner, J. "Donkeys as Livestock Guards." *Mother Earth News*, October 18, 2013. www.motherearthnews.com/homesteading-and-livestock/guard-donkey-zbcz1310.

Dolan, J. M. "The Quagga Commemorated." *Zoonooz* 56, no. 8 (1983): 13–15.

Dorst, J. "Notice Sur Les Spécimens Naturalisés De Mammifères Éteints Existant Dans Les Collections Du Muséum." *Bulletin du Muséum National d'Histoire Naturelle* 24, no. 1 (1952): 63–79.

Dubow, S. "Afrikaner Nationalism, Apartheid and the Conceptualization of 'Race'." *The Journal of African History* 33, no. 2 (1992): 209–37.

Duncan, P., and C. P. Groves. "Genus Equus: Asses, Zebras." In *Mammals of Africa*, Vol 5. Edited by J. Kingdon and M. Hoffmann, 412–13. London: Bloomsbury Publishing, 2013.

Dutton, P., and S. Dutton. "Tragedy in the Making: Mozambique's Endemic Selous Zebra Destined to Join the Extinct Quagga." *African Wildlife* 54, no. 3 (2000): 27.

Edwards, G. *Gleanings of Natural History*. London: Royal College of Physicians, 1758.

Edwards, J. "Live Quagga Photographs." *The Mane* 4 (1997): 7.

Egri, A., M. Blaho, G. Kriska, R. Farkas, M. Gyurkovszky, S. Akesson, and G. Horvath. "Polarotactic Tabanids Find Striped Patterns with Brightness and/or Polarization Modulation Least Attractive: An Advantage of Zebra Stripes." *Journal of Experimental Biology* 215, no. 5 (2012): 736–45.

Eisenmann, V., and J. S. Brink. "Koffiefontein Quaggas and True Cape Quaggas: The Importance of Basic Skull Morphology." *South African Journal of Science* 96 (2000): 529–33.

Ellis, A. B. *South African Sketches*. London: Chapman and Hall Limited, 1887.

Eloff, G. "The Passing of the True Quagga and the Little Klibbolikhonnifontein Burchell's Zebra. Which Is to Be Next, the Cape Mountain Zebra or Wahlberg's Zebra of Zululand?" *Tydskrif vir Natuurwetenskappe* 6 (1966): 193–207.

Elphick, R. *Kraal and Castle: Khoikhoi and the Founding of White South Africa*. New Haven, CT: Yale University Press, 1977.

Endangered Wildlife Trust. "2016 Red List of Mammals of South Africa, Lesotho and Swaziland." 2016. https://endangeredwildlifetrust.wordpress.com/2017/01/26/2016-red-list-of-mammals-of-south-africa-lesotho-and-swaziland/.

——. "Perspective on the Intensive Breeding of Wildlife Species with Particular Reference to Selective Breeding for Colour Variants." 2016. www.ewt.org.za/wp-content/uploads/2019/04/EWT-Perspective-on-Wildlife-Intensive-Breeding-and-Colour-Variation-March-2016.pdf.

Enright, D. J. *Selected Poems*. London: Chatto & Windus, 1968.

Eschner, K. "Those Little Birds on the Backs of Rhinos Actually Drink Blood." *Smithsonian*, September 22, 2017. www.smithsonianmag.com/smart-news/those-little-birds-backs-rhinos-actually-drink-blood-180964912/.

Estes, R. *The Behaviour Guide to African Mammals*. Berkeley: The University of California Press, 1991.

Ewart, J. C. *The Penycuik Experiments*. London: A. and C. Black, 1899.

——, T. Constable, and A. Constable. *Guide to the Zebra Hybrids etc. On Exhibition at the Royal Agricultural Society's Show, York: Together with a Description of Zebras, Hybrids, Telegony, etc.* Edinburgh: Printed by T. and A. Constable, Printer to Her Majesty, 1900.

Faith, J. T. "Palaeozoological Insights into Management Options for a Threatened Mammal: Southern Africa's Cape Mountain Zebra (Equus zebra zebra)." *Diversity and Distributions* 18, no. 5 (2012): 438–47.

Feely, J. M. "IsiXhosa Names of South African Land Mammals." *Africa Zoology* 44, no. 2 (2009): 141–50.

"The Last Quagga in Transkei." *African Wildlife* 43, no. 5 (1989): 245.

Fennessy, J., T. Bidon, F. Reuss, V. Kumar, P. Elkan, M. A. Nilsson, . . . A. Janke. "Multi-Locus Analyses Reveal Four Giraffe Species Instead of One." *Current Biology* 26, no. 18 (2016): 2543–49.

Ferreira, J. "Quagga Rebreeding a Success Story." *Farmer's Weekly* (South Africa), no. 14011 (2014): 50–53.

Ferris, M. A. "Parks Body Joins Quest to 'Re-Breed' the Quagga." *The Star*. July 3, 2000.

Fischhoff, I. R., S. R. Sundaresan, J. Cordingley, H. M. Larkin, M. Sellier, and D. I. Rubenstein. "Social Relationships and Reproductive State Influence Leadership Roles in Movements of Plains Zebra, Equus burchellii." *Animal Behaviour* 73, no. 5 (2007): 825–31.

Fitzsimons, F. W. *The Natural History of South Africa: Mammals*, Vol. 3. London: Longmans, Green & Co., 1920.

Flint, A. *A Textbook of Human Physiology*, 4th ed. New York: D. Appleton & Co., 1888.

Fock, G. J., and A. McGregor, "Two Rock Engravings from South Africa in the British Museum." *Man* 65 (1965): 194–95.

Ford, W., S. Katz, and D. Kazanjian. *Walton Ford: Tigers of Wrath, Horses of Instruction*. New York: H. N. Abrams, 2002.

Freeman, C. "Ending Extinction: The Quagga, the Thylacine, and the 'Smart Human'." In *Leonardo's Choice*. Edited by C. Gigliotti, 235–56. New York: Springer, 2009.

    "Imaging Extinction: Disclosure and Revision in Photographs of the Thylacine (Tasmanian Tiger)." *Society and Animals* 15, no. 3 (2007): 241–56.

Friedel, G. "The Story of 'Howie': Umgeni Valley's Quagga-Like Plains Zebra." *African Wildlife* 54, no. 2 (2000): 9–10.

Fritsch, G. *Drei Jahre in Süd-Afrika: Reiseskizzen Nach Notizen Des Tagebuchs Zusammengestellt: Mit Zahlreichen Ill*, Vol. 1. Breslau, Germany: Hirt, 1868.

Frost, R. "The Oven Bird." *Poetry Foundation*. www.poetryfoundation.org/poems/44269/the-oven-bird.

Gall, D. M. "This Most Elegant of Quadrupeds: Illustrations of the Quagga." *Discovery, Journal of the Peabody Museum of Natural History* 15, no. 2 (1980): 44–50.

Gatti, A. *Here Is the Veld*. New York: C. Scribner's Sons, 1948.

Gaywood, M. J. "Reintroducing the Eurasian Beaver *Castor fiber* to Scotland." *Mammal Review* 48, no. 1 (2018): 48–61.

Gens, E. *Promenade Au Jardin Zoologique D'Anvers*. Anvers, Belgium: J.-E. Buschmann, 1861. www.biodiversity library.org/item/127664.

Gordon, R. J. "Equus quagga quagga (Quagga)." *Robert Jacob Gordon: His Verbal and Visual Descriptions of South Africa*. Edited by Rijksmuseum, 1777. https://robertjacobgordon.nl/drawings/rp-t-1914-17-190.

    The Gordon African Collection (RP-T-1914-17-190). Amsterdam: Rijksmuseum, 2016.

    "Travel Journals. Second Journey." *Robert Jacob Gordon: His Verbal and Visual Descriptions of South Africa*. Edited by Rijksmuseum, 1777. https://robertjacobgordon.nl/writings-and-drawings.

Gordon, R. J., P. E. Raper, and M. Boucher. *Robert Jacob Gordon: Cape Travels: 1777 to 1786*, Houghton, South Africa: Brenthurst, 1988.

Gordon-Cumming, R. G. *A Hunter's Life among Lions, Elephants, and Other Wild Animals of South Africa*. New York: Derby & Jackson, 1856.

Gosling, L. M., J. Muntifering, H. Kolberg, K. Uiseb, and S. R. B. King. "Equus zebra Ssp. hartmannae." *The IUCN Red List of Threatened Species*, 2019. www.iucnredlist.org/species/7958/45171819.

Gould, G. M., and W. L. Pyle. *Anomalies and Curiosities of Medicine*. London: W. B. Saunders, 1901.

Gould, S. J. "The Return of Hopeful Monsters." *Natural History* 86, no. 6 (1977): 22–30.

Wonderful Life: The Burgess Shale and the Nature of History. New York: W. W. Norton, 1989.

Govindasamy, V. "Stars and Stripes by SA Animators." 2013. www.filmcontact.com/africa/south-africa/stars-and-stripes-sa-animators.

Graff, D. "Shifting Sands: An Interest-Relative Theory of Vagueness." *Philosophical Topics* 28, no. 1 (2000): 45–81.

Gray, J. E. "A Revision of the Family Equidae." *Zoological Journal* 1, no. 2 (1825): 241–48.

Gray, J. E., B. W. Hawkins, and E. H. S. Derby. *Gleanings from the Menagerie and Aviary at Knowsley Hall*. Knowsley: Printed for Private Distribution, 1850.

Green, L. G. *Karoo*. Cape Town: H. Timmins, 1955.

Green, T. *The Quagga*. Milwaukee, WI: Gareth Stevens Publishers, 1996.

Greenberg, J. *A Feathered River across the Sky: The Passenger Pigeon's Flight to Extinction*. New York: Bloomsbury USA, 2014.

Greig, J. C. "1883–1983: Centennial of the Extinction of the Quagga." *African Wildlife* 37, no. 4 (1983): 146–48.

"The Lesson of the Quagga." *African Wildlife* 37, no. 4 (1983): 134–35.

"The Origin of the Name 'Quagga'." *African Wildlife* 37, no. 4 (1983): 149–51.

"The Quagga and the Taxidermist's Art." *African Wildlife* 37, no. 4 (1983): 140–41.

Grigson, C. *Menagerie: A History of Exotic Animals in England*. Oxford: Oxford University Press, 2016.

Groves, C. P. "Taxonomy of Living Equidae." In *Equids: Zebras, Asses and Horses. Status Survey and Conservation Action Plan*. Edited by P. Moehlman, 94–107. Gland, Switzerland: The World Conservation Union (IUCN), Equid Specialist Group, 2002.

"Was the Quagga a Species or a Sub-Species?" *African Wildlife* 39, no. 3 (1985): 106–7.

Groves, C. P., and C. H. Bell, "New Investigations on the Taxonomy of the Zebras Genus Equus, Subgenus Hippotigris." *Mammalian Biology* 69, no. 3 (2004): 182–96.

Groves, C. P., and D. Happold. "Classification: A Mammalian Perspective." In *Mammals of Africa, Volume I: Introductory Chapters and Afrotheria*. Edited by J. Kingdon, D. Happold, M. Hoffmann, T. Butynski, M. Happold, and J. Kalina, 101–8. London: Bloomsbury Publishing, 2013.

Grubb, P. "Types and Type Localities of Ungulates Named from Southern Africa." *Koedoe: African Protected Area Conservation and Science* 42, no. 2 (1999): 13–45.

Gurdon, J. B. "Transplanted Nuclei and Cell Differentiation." *Scientific American* 219, no. 6 (1968): 24–35.

Hack, M. A., R. East, and D. I. Rubenstein. *Status and Action Plan for the Plains Zebra (Equus burchellii)*. Edited by P. D. Moehlman. https://portals.iucn.org/library/efiles/documents/2002-043.pdf.

Harley, E. H., C. Lardner, M. Gregor, B. Wooding, and M. H. Knight. "The Restoration of the Quagga; 24 Years of Selective Breeding." In *Restoration of Endangered and Extinct Animals*. Edited by R. Slomski. Poznan, Poland: Poznan University of Life Sciences Press, 2010.

Harley, E. H., M. de Waal, S. Murray, and C. O'Ryan. "Comparison of Whole Mitochondrial Genome Sequences of Northern and Southern White Rhinoceroses (Ceratotherium simum): The Conservation Consequences of Species Definitions." *Conservation Genetics* 17, no. 6 (2016): 1285–91.

Harley, E. H. and B. Wooding. "Quagga Project FAQ." *The Quagga Project*, n.d. https://quaggaproject.org/faq/.

Harris, P. "Scientists Trying to Revive Extinct Kin of Zebras." *Los Angeles Times.* August 30, 1998: 8.

Harris, W. C. "*Aigocerus niger*. The Sable Antelope." *Proceedings of the Zoological Society of London* 4 (1838): 1–3.

"Description of a New Species of Antelope." *The Transactions of the Zoological Society of London* 2 (1839): 213–15.

Narrative of an Expedition into Southern Africa, during the Years 1836, and 1837. Bombay: American Mission, 1838.

Portraits of the Game & Wild Animals of Southern Africa. London: Pickering, 1840.

The Wild Sports of Southern Africa. London: John Murray, 1839.

Hassan, H. Y., A. van Erp, M. Jaeger, H. Tahir, M. Oosting, L. A. Joosten, and M. G. Netea. "Genetic Diversity of Lactase Persistence in East African Populations." *BMC Research Notes* 9, no. 8 (2016).

Hawkins, W. "Two Quaggas Belonging to Lord Knowsley Kept at His Estate near Liverpool, England." 1850. Lithograph. https://biodiversitylibrary.org/page/47595182.

Hayward, M. W., S. Edwards, B. A. Fancourt, J. D. C. Linnell, and E. B. Nilsen. "Top-Down Control of Ecosystems and the Case for Rewilding: Does It All Add Up?" In *Rewilding*. Edited by N. Pettorelli, S. M. Durant, and J. T. Du Toit, 325–54. Cambridge: Cambridge University Press, 2019.

Hayward, M. W., and G. I. H. Kerley. "Prey Preferences of the Lion (Panthera leo)." *Journal of Zoology* 267, no. 3 (2005): 309–22.

Heck, H. "The Breeding-Back of the Aurochs." *Oryx* 1, no. 3 (1951): 117–22.

Heck, L. *Animal Safari; Big Game in South West Africa.* London: Methuen & Co., 1956.

Animals, My Adventure. London: Methuen & Co., 1954.

Hedrick, P. W. "Population Genetics of Malaria Resistance in Humans." *Heredity* 107, no. 4 (2011): 283–304.

Hendricks, M. D. "Remaking |Xam Narratives in a Post-Apartheid South Africa." Mini-Thesis, University of the Western Cape, 2010.

Hennessy, E. "Saving Species: The Co-Evolution of Tortoise Taxonomy and Conservation in the Galapagos Islands." *Environmental History* 25, no. 2 (2020): 263–86.

Hewitt, R. *Structure, Meaning and Ritual in the Narratives of the Southern San.* 2018. www.cambridge.org/core/product/identifier/9781868147038/type/BOOK.

Heywood, P. "The Micro-Politics of Macromolecules in the Taxonomy and Restoration of Quaggas." *Kronos,* no. 41 (2015): 314–37.

"The Quagga and Science: What Does the Future Hold for This Extinct Zebra?" *Perspectives in Biology and Medicine* 56, no. 1 (2013): 53–64.

"Sexual Dimorphism of Body Size in Taxidermy Specimens of Equus quagga quagga Boddaert (Equidae)." *Journal of Natural History* 53, no. 45 (2019): 2757–61.

"Ways of Seeing Nonhuman Animals. Some Likened Zebras to Horses, Others to Asses." *Society and Animals* (2020), published online ahead of print. doi: https://doi.org/10.1163/15685306-BJA10027.

Higuchi, R. G., B. Bowman, M. Freiberger, O. A. Ryder, and A. C. Wilson. "DNA Sequences from the Quagga, an Extinct Member of the Horse Family." *Nature* 312, no. 5991 (1984): 282–84.

Higuchi, R. G., L. A. Wrischnick, E. Oakes, M. George, B. Tong, and A. Wilson. "Mitochondrial DNA of the Extinct Quagga: Relatedness and Extent of Postmortem Change." *Journal of Molecular Evolution* 25, no. 4 (1987): 283–87.

Hoare, D. B., L. Mucina, M. C. Rutherford, J. H. J. Vlok, D. I. W. Euston-Brown, A. R. Palmer, ... R. A. Ward. "Albany Thicket Biome." In *The Vegetation of South Africa, Lesotho and Swaziland.* Edited by L. Mucina and M. C. Rutherford, 540–67. Pretoria: South African National Biodiversity Institute, 2006.

Hoffman, M. T., B. Cousins, T. Meyer, A. Petersen, and H. Hendricks. "Historical and Contemporary Land Use and the Desertification of the Karoo." In *The Karoo: Ecological Patterns and Processes.* Edited by W. R. J. Dean and S. J. Milton, 257–73. Cambridge: Cambridge University Press, 1999.

Hofreiter, M. "Palaeogenomics." *Comptes Rendus – Palevol* 7, no. 2–3 (2008): 113–24.

Hopkins, G. M. "Binsey Poplars." *Poetry Foundation.* 1879. www.poetryfoundation.org/poems/44390/binsey-poplars.

How, M. J., D. Gonzales, A. Irwin, and T. Caro. "Zebra Stripes, Tabanid Biting Flies and the Aperture Effect." *Proceedings of the Royal Society B: Biological Sciences* 287, no. 1933 (2020): 1521.

How, M. J., and J. M. Zanker. "Motion Camouflage Induced by Zebra Stripes." *Zoology* 117, no. 3 (2014): 163–70.

Hrabar, H., C. Birss, D. Peinke, P. Novellie, and G Kerley. "Equus zebra Ssp. zebra." *The IUCN Red List of Threatened Species,* 2019. www.iucnredlist.org/species/7959/45171853, last accessed November 17, 2020.

Hrabar, H., and G. Kerley. *Selective Breeding in the Quagga Breeding Program: The Effect of Translocations and Inbreeding on Plains Zebra Reproduction.* Port Elizabeth, South Africa: Center for African Conservation Ecology, 2013. www.semanticscholar.org/paper/Selective-breeding-in-the-Quagga-Breeding-Program-%E2%80%93-Hrabar-Kerley/806be4c33ecea04b9f84cede3ad202b02efa4fc4.

Hrabar, H., and G. I. H. Kerley. "Conservation Goals for the Cape Mountain Zebra Equus zebra zebra: Security in Numbers?" *Oryx* 47, no. 3 (2013): 403–9.

Huber, W. "Das Münchner Quagga: Eine Zoologische Rarität." *Spixiana* 17 (1992): 155–60.

"Documentation of the Five Known Photographs of a Living Quagga, Equus quagga quagga Gmelin, 1788 (Mammals, Perissodactyla, Equidae)." *Spixiana* 17, no. 2 (1994): 193–99.

Hudson, A. L. S. "Infantry Weapons in South Africa, 1652–1881." *Scientia Militaria, South African Journal of Military Studies* 10, no. 2 (1980): 48–53.

Hughes, A. E., J. Troscianko, and M. Stevens. "Motion Dazzle and the Effects of Target Patterning on Capture Success." *BMC Evolutionary Biology* 14, no. 1 (2014): 201.

Hulley, B. "Farewell to the Quagga Pioneer." Cape Towner (Cape Town). March 2, 2006, 11.

Isenberg, A. C. *The Destruction of the Bison: An Environmental History, 1750–1920.* Cambridge: Cambridge University Press, 2000.

Itano, N. "S. Africa's Quagga Saga: Righting a Past Wrong." *Christian Science Monitor.* October 1, 2003, 7.

Jackson, A. de J. *Manna in the Desert: A Revelation of the Great Karroo.* Johannesburg: Christian Literature Depot, 1919.

Jacobs, N. J. *Environment, Power, and Justice: A South African History.* New York: Cambridge University Press, 2003.

"The Great Bophuthatswana Donkey Massacre." *American Historical Review* 106, no. 2 (2001): 485–507.

JAG.TV, WILD &. Wildswinkel Boland Veiling. YouTube, 2017.

Janis, C. "The Evolutionary Strategy of the Equidae and the Origins of Rumen and Cecal Digestion." *Evolution* 30 (1976): 757–74.

Jarvis, B. "Paper Tiger. Could a Global Icon of Extinction Still Be Alive?" *The New Yorker.* July 2, 2018.

Johns, D. "History of Rewilding: Ideas and Practice." In *Rewilding.* Edited by N. Pettorelli, S. M. Durant, and J. T. du Toit, 12–33. Cambridge: Cambridge University Press, 2019.

Johnson, L. M. "A Clone of the Endangered Przewalski's Horse Is Born of DNA Saved for 40 Years." 2020. www.cnn.com/2020/09/12/us/cloned-przewalskis-horse-trnd/index.html?fbclid=IwAR2dpDqF7m1 kqig lHYJ51IOr-tpkaPJwUqKC g7f3njOC1pfBWHYE7r1jPs4.

Jónsson, H., M. Schubert, A. Seguin-Orlando, A. Ginolhac, L. Petersen, M. Fumagalli, … L. Orlando. "Speciation with Gene Flow in Equids Despite Extensive Chromosomal Plasticity." *Proceedings of the National Academy of Sciences (USA)* 111, no. 52 (2014): 18655–60.

Jordan, B. "Karoo Welcomes Quagga Back from the Dead." Sunday Times, September 27, 1998.

Kellert, S. R. *The Value of Life: Biological Diversity and Human Society.* Washington, DC: Island Press, 1997.

Keynes, S. *Quentin Keynes (1921–2003): Explorer, Film-Maker, Lecturer, and Book-Collector.* Cambridge: Cambridge University Press, 2004.

Kimura, M. *The Neutral Theory of Molecular Evolution*. Cambridge: Cambridge University Press, 1983.

King, J. M. "A Field Guide to the Reproduction of the Grant's Zebra and Grevy's Zebra." *AJE African Journal of Ecology* 3, no. 1 (1965): 99–117.

King, S. R. B., and P. D. Moehlman. "Equus quagga." *The IUCN Red List of Threatened Species*, 2016. www.iucnredlist.org/species/41013/45172424.

Kingdon, J. *The Kingdon Field Guide to African Mammals*. London: Bloomsbury, 2015.

"Subgenus Hippotigris." In *Mammals of Africa. Vol. 5, Carnivores, Pangolins, Equids and Rhinoceroses*. Edited by J. Kingdon and M. Hoffmann, 417–21. London: Bloomsbury Publishing, 2013.

Kipling, R. *How the Leopard Got His Spots*. The Project Guttenberg EBook of Just So Stories, by Rudyard Kipling, 1902.

Kivinen, S., P. Nummi, and T. Kumpula. "Beaver-Induced Spatiotemporal Patch Dynamics Affect Landscape-Level Environmental Heterogeneity." *Environmental Research Letters* 15, no. 9 (2020): 094065.

Klak, C., G. Reeves, and T. Hedderson. "Unmatched Tempo of Evolution in Southern African Semi-Desert Ice Plants." *Nature* 427, no. 6969 (2004): 63–65.

Klein, R. G. "A Preliminary Report on the Larger Mammals from the Boomplaas Stone Age Cave Site, Cango Valley, Oudtshoorn District, South Africa." *The South African Archaeological Bulletin* 33, no. 127 (1978): 66–75.

Klein, R. G., and K. Cruz-Uribe. "Craniometry of the Genus *Equus* and the Taxonomic Affinities of the Extinct South African Quagga." *South African Journal of Science* 95, no. 2 (1999): 81–86.

Klingel, H. "*Equus quagga* Plains Zebra (Common Zebra)." In *Mammals of Africa. Vol. 5, Carnivores, Pangolins, Equids and Rhinoceroses*. Edited by J. Kingdon and M. Hoffmann, 428–37. London: Bloomsbury Publishing, 2013.

Kolbert, E. "Recall of the Wild: The Quest to Engineer a World before Humans." The New Yorker. December 24, 2012, 50–60.

La Cépède, E. de. *La Ménagerie du Muséum National D'Histoire Naturelle, Ou Description et Histoire des Animaux*. Paris: Miger, Graveur et Renouard, 1804.

Landman, M., and G. I. H. Kerley. "Dietary Shifts: Do Grazers Become Browsers in the Thicket Biome?" *Koedoe: African Protected Area Conservation and Science* 44, no. 1 (2001): 31–36.

Larison, B., R. J. Harrigan, H. A. Thomassen, D. I. Rubenstein, A. M. Chan-Golston, E. Li, and T. B. Smith. "How the Zebra Got Its Stripes: A Problem with Too Many Solutions." *Royal Society of Open Science* 2 (2015): 140452.

Larison, B., C. B. Kaelin, R. J. Harrigan, C. Henegar, D. Rubenstein, T. B. Smith, ... O. Aschenborn. "Population Structure, Inbreeding and Stripe Pattern Abnormalities in Plains Zebras." *Molecular Ecology* 30, no. 2 (2021): 379–90.

Larkin, N. R. and L. B. Porro. "Three Legs Good, Four Legs Better: Making a Quagga Whole Again with 3d Printing." *Collection Forum* 30, no. 1–2 (2016): 73–84.

Lategan, F. V. and L. Potgieter. *Die Boer se Roer Tot Vandag: Die Ontwikkeling van die Vuurwapen in Suider-Afrika*. Kaapstad: Tafelberg, 1982.

"Le Couagga De Louis XVI." n.d. www.mnhn.fr/fr/collections/ensembles-collec tions/vertebres/mammiferes/couagga-louis-xvi.

Le Fanu, W. *A Catalogue of the Portraits and Other Paintings, Drawings and Sculpture in the Royal College of Surgeons of England.* Edinburgh: E. & S. Livingstone, 1960.

Leonard, J. A., N. Rohland, S. Glaberman, R. C. Rleischer, A. Caccaone, and M. Hofreiter. "A Rapid Loss of Stripes: The Evolutionary History of the Extinct Quagga." *Biology Letters* 1, no. 3 (2005): 291–95.

Lichtenstein, H. *Travels in Southern Africa in the Years 1803, 1804, 1805 and 1806.* 2 vols. London: H. Colburn, 1812–15.

Limburg, K. E., V. A. Luzadis, M. Ramsey, K. L. Schulz, and C. M. Mayer. "The Good, the Bad, and the Algae: Perceiving Ecosystem Services and Disservices Generated by Zebra and Quagga Mussels." *Journal of Great Lakes Research* 36, no. 1 (2010): 86–92.

Lindsay, S. "Elegy for the Quagga." *Poetry* 193, no. 1 (2008): 6.

von Linné, C., and J. Gmelin. *Caroli a Linné ... Systema Naturæ Per Regna Tria Naturae,: Secundum Classes, Ordines, Genera, Species, Cum Characteribus, Differentiis, Synonymis, Locis.* Lipsiæ: J. B. Delamolliere, 1789.

Linnell, J. D. C., and C. R. Jackson. "Bringing Back Large Carnivores to Rewild Landscapes." In *Rewilding.* Edited by N. Pettorelli, S. M. Durant, and J. T. du Toit, 248–79. Cambridge: Cambridge University Press, 2019.

Lorenzen, E. D., P. Arctander, and H. R. Siegismund. "High Variation and Very Low Differentiation in Wide Ranging Plains Zebra (*Equus quagga*): Insights from mtDNA and Microsatellites." *MEC Molecular Ecology* 17, no. 12 (2008): 2812–24.

Lorimer, J. *Wildlife in the Anthropocene: Conservation after Nature.* Minneapolis: University of Minnesota Press, 2015.

Lowenstein, J. M. "Half-Striped Quagga Was a Plains Zebra." New Scientist, July 18, 1985, 27.

Lowenstein, J. M., and O. A. Ryder "Immunological Systematics of the Extinct Quagga (Equidae)." *Experientia* 41, no. 9 (1985): 1192–93.

Lundholm, B. "Is Rebirth of the Quagga Possible?" *African Wildlife* 5, no. 3 (1951): 209–12.

"A Skull of the True Quagga (Equus quagga) in the Collection of the Transvaal Museum." *South African Journal of Science* 47, no. 11 (1951): 307–12.

Lydekker, R. *Catalogue of the Ungulate Mammals in the British Museum (Natural History).* London: Printed by order of the Trustees of the British Museum, 1916.

The Game Animals of Africa. London: Rowland Ward, 1908.

"Note on the Skull and Markings of the Quagga." *Proceedings of the Zoological Society of London* 74, no. 2 (1904): 426–31.

MacClintock, D. *A Natural History of Zebras.* New York: Scribner, 1976.

"Professor Marsh's Quagga Mare." *Discovery, Journal of the Peabody Museum of Natural History* 15, no. 2 (1980): 34–43.

MacKenzie, J. M. *The Empire of Nature: Hunting, Conservation and British Imperialism.* Manchester: Manchester University Press, 1988.

Macnab, R. "The Quagga's World." In *Mantis Poets*. Edited by J. Cope, 7. Cape Town: David Philip, 1981.

"Winged Quagga." In *Mantis Poets*. Edited by J. Cope, 4. Cape Town: David Philip, 1981.

Malan, B. D., and H. B. S. Cooke. *A Preliminary Account of the Wonderwerk Cave, Kuruman District. Section 2: The Fossil Remains*. Cape Town: Griffiths, 1940.

Manganiello, C. J. "From a Howling Wilderness to Howling Safaris: Science, Policy and Red Wolves in the American South." *Journal of the History of Biology* 42, no. 2 (2009): 325–59.

Maréchal, N. "Quagga." 1793. https://commons.wikimedia.org/wiki/File:Quagga .jpg, last accessed November 10, 2021.

Masubelele, M. L., M. T. Hoffman, W. Bond, and P. Burdett. "Vegetation Change (1988–2010) in Camdeboo National Park (South Africa), Using Fixed-Point Photo Monitoring: The Role of Herbivory and Climate." *Koedoe: African Protected Area Conservation and Science* 55, no. 1 (2013): 1–16.

Max, D. T. "Can You Revive an Extinct Animal?" New York Times. January 1, 2006.

Mayr, E. "What Is a Species, and What Is Not?" *Philosophy of Science* 63, no. 2 (1996): 262–77.

McHorse, B. K., A. A. Biewener, and S. E. Pierce. "Mechanics of Evolutionary Digit Reduction in Fossil Horses (Equidae)." *Proceedings of the Royal Society of Biological Sciences* 284, no. 1861 (2017): 1174.

McNeil, D. G. "Brave Quest of Africa Hunt: Bringing Back Extinct Quagga." New York Times. September 16, 1997.

Meadows, M. E., and M. K. Watkeys. "Paleoenvironments." In *The Karoo: Ecological Patterns and Processes*. Edited by W. R. J. Dean and S. J. Milton, 27–41. Cambridge: Cambridge University Press, 1999.

Mendel, G. "Versuche Uber Pflanzen-Hybriden." *Verhandlungen des naturforschenden Vereins Brünn* 4 (1865): 3–47.

Mentzel, O. F. *A Geographical and Topographical Description of the Cape of Good Hope*, Vol. 3. Cape Town: Van Riebeeck Society, 1944.

Meuwissen, T., B. Hayes, and M. Goddard. "Genomic Selection: A Paradigm Shift in Animal Breeding." *Animal Frontiers* 6, no. 1 (2016): 6–14.

Midgley, G. F., and F. van der Heyden. "Form and Function in Perennial Plants." In *The Karoo: Ecological Patterns and Processes*. Edited by W. R. J. Dean and S. J. Milton, 91–106. Cambridge: Cambridge University Press, 1999.

Miller, J. M., M. C. Quinzin, N. Poulakakis, J. P. Gibbs, L. B. Beheregaray, R. C. Garrick, ... A. Caccone. "Identification of Genetically Important Individuals of the Rediscovered Floreana Galápagos Giant Tortoise (Chelonoidis elephantopus) Provides Founders for Species Restoration Program." *Scientific Reports* 7, no. 1 (2017): 11471.

Mills, E. L., G. Rosenberg, A. P. Spidle, M. Ludyanskiy, Y. Pligin, and B. May. "A Review of the Biology and Ecology of the Quagga Mussel (*Dreissena bugensis*), a Second Species of Freshwater Dreissenid Introduced to North America." *American Zoologist* 36, no. 3 (1996): 271–86.

Milton, S. J., R. A. G. Davies, and G. I. H. Kerley. "Population Level Dynamics." In *The Karoo: Ecological Patterns and Processes*. Edited by W. R. J. Dean and S. J. Milton, 183–207. Cambridge: Cambridge University Press, 1999.

Morrison, R., and S. C. Watson. *Seven Moon Circus*. San Diego, CA: Ringleader Books, 2013.

Morton, E. O. "A Communication of a Singular Fact in Natural History." *Philosophical Transactions of the Royal Society of London* 111 (1821): 20–22.

Mucina, L., A. le Roux, M. C. Rutherford, U. Schmiedel, K. Esler, L. Powrie, ... N. Jürgens. "Succulent Karoo Biome." In *The Vegetation of South Africa, Lesotho and Swaziland*. Edited by L. Mucina and M. C. Rutherford. Strelitzia 19, 220–99. Pretoria: South African National Biodiversity Institute, 2006.

Mungall, E. R. [née Cary]. "Extinction: The Quagga Mapped through Time." *Environmental Awareness* 10, no. 2 (1987): 53–62.

Naidoo, R., M. J. Chase, K. Landen, P. Beytell, P. du Preez, G. Stuart-Hill, and R. Taylor. "A Newly Discovered Wildlife Migration in Namibia and Botswana Is the Longest in Africa." *Oryx* 50, no. 1 (2014): 138–46.

Navarro, L. M., and H. M. Pereira. "Rewilding Abandoned Landscapes in Europe." *Ecosystems* 15, no. 6 (2012): 900–12.

Neuhaus, P., and K. E. Ruckstuhl. "The Link between Sexual Dimorphism, Activity Budgets, and Group Cohesion: The Case of the Plains Zebra (Equus burchelli)." *Canadian Journal of Zoology* 80 (2002): 1437–41.

Novak, B. J. "De-Extinction." *Genes* 9, no. 11 (2018): 548.

Novellie, P., M. Lindeque, P. Lindeque, P. Lloyd, and J. Koen. "Status and Action Plan for the Mountain Zebra (Equus zebra)." In *Equids: Zebras, Asses and Horses: Status Survey and Conservation Action Plan*. Edited by P. Moehlman, 28–42. Gland, Switzerland: International Union for Conservation of Nature (IUCN), 2002.

Nunan, E. "In Their True Colors: Developing New Methods for Recoloring Faded Taxidermy." n.d. http://intheirtrue colors.wordpress.com.

Orlando, L., A. Ginolhac, G. Zhang, D. Froese, A. Albrechtsen, M. Stiller, ... E. Willerslev. "Recalibrating Equus Evolution Using the Genome Sequence of an Early Middle Pleistocene Horse." *Nature* 499, no. 7456 (2013): 74–78.

Orlando, L., J. L. Metcalf, M. T. Alberdi, M. Telles-Antunes, D. Bonjean, M. Otte, ... F. Morello. "Revising the Recent Evolutionary History of Equids Using Ancient DNA." *Proceedings of the National Academy of Sciences (USA)* 106, no. 51 (2009): 21754–59.

Ottenburghs, J., P. van Hooft, S. E. van Wieren, R. C. Ydenberg, and H. H. T. Prins. "Hybridization in Geese: A Review." *Frontiers in Zoology* 13, no. 1 (2016): 20.

Paddle, R. N. *The Last Tasmanian Tiger: The History and Extinction of the Thylacine*. Cambridge: Cambridge University Press, 2002.

Page, T., and C. Hancock. "Zebra Cousin Went Extinct 100 Years Ago. Now, It's Back." January 27, 2016. https://edition.cnn.com/2016/01/25/africa/quagga-project-zebra-conservation-extinct-south-africa/.

Palmer, A. R., P. A. Novellie, and J. W. Lloyd. "Community Patterns and Dynamics." In *The Karoo: Ecological Patterns and Processes*. Edited by W. R. J. Dean and S. J. Milton, 208–23. Cambridge: Cambridge University Press, 1999.

Parsons, R., C. Aldous-Mycock, and M. R. Perrin. "A Genetic Index for Stripe-Pattern Reduction in the Zebra: The Quagga Project." *Southern African Journal of Wildlife Research* 37, no. 2 (2007): 105–16.

Pearson, K., and F. Galton. *The Life, Letters and Labours of Francis Galton*, Vol. 2. Cambridge: Cambridge University Press, 1924.

Pedersen, C. T., A. Albrechtsen, P. D. Etter, E. A. Johnson, L. Orlando, … R. Heller. "A Southern African Origin and Cryptic Structure in the Highly Mobile Plains Zebra." *Nature Ecology & Evolution* 2, no. 3 (2018): 491–98.

Penn, N. *The Forgotten Frontier: Colonist and Khoisan on the Cape's Northern Frontier in the 18th Century*. Athens: Ohio University Press, 2007.

Penzhorn, B. "*Equus zebra* Mountain Zebra." In *Mammals of Africa. Vol. 5, Carnivores, Pangolins, Equids and Rhinoceroses*. Edited by J. Kingdon and M. Hoffmann, 438–43. London: Bloomsbury Publishing, 2013.

"Equus zebra." *The American Society of Mammalogists* 314 (1988): 1–7.

Pettorelli, N., S. M. Durant, and J. T. du Toit. *Rewilding*. Cambridge: Cambridge University Press, 2019.

Phillips, M. K., V. G. Henry, and B. T. Kelly. "Restoration of the Red Wolf." *USGS Northern Prairie Wildlife Research Center* 319 (2003): 272–88.

Plotz, R., and W. Linklater. "Red-Billed Oxpeckers Really Do Increase Predator Awareness in Black Rhinoceros." *13th International Behavioral Ecology Congress* (2010).

Plumb, C. "The Queen's Ass." In *The Afterlives of Animals: A Museum Menagerie*. Edited by S. J. M. M. Alberti, 17–36. Charlottesville: University of Virginia Press, 2011.

Plumb, C., and S. Shaw. *Zebra*. London: Reaktion Books, 2018.

Pocock, R. I. "The Cape Colony Quaggas." *Journal of Natural History* 14, no. 83 (1904): 313–28.

"The Coloration of the Quaggas." *Nature* 68, no. 1763 (1903): 356–57.

"On the Preorbital Pit in the Skulls of Domestic Horses and Quaggas." *Annals and Magazine of Natural History* 15, no. 89 (1905): 516–18.

Poinar, G. "Ancient DNA." *American Scientist* 87 (1999): 446–57.

Popkin, G. "Can Genetic Engineering Bring Back the American Chestnut?" New York Times. April 30, 2020.

Powell, E. "Zebra Meat: Exotic and Lean – but Does It Taste Good?" The Independent. July 26, 2014.

Pringle, T. *African Sketches*. London: Edward Moxon, 1834.

Narrative of a Residence in South Africa. Edited by J. Conder. London: Edward Moxon, 1835.

Protheroe, E. *New Illustrated Natural History of the World*. Garden City, NY: Garden City Publishing Co., 1940.

Radloff, F. G. T., L. Mucina, W. J. Bond, and P. J. le Roux. "Strontium Isotope Analyses of Large Herbivore Habitat Use in the Cape Fynbos Region of South Africa." *Oecologia* 164, no. 2 (2010): 567–78.

Rainier, M. "Some Quagga Place-Names and Some Misconceptions." *African Journal of Wildlife Research* 38, no. 1 (1984): 126.

Rau, R. E. "Additions to the Revised List of Preserved Material of the Extinct Cape Colony Quagga and Notes on the Relationship and Distribution of Southern Plains Zebras." *Annals of the South African Museum* 77, no. 2 (1978): 27–45.

"Colouration Abnormalities in the Plains Zebra." *African Wildlife* 57, no. 2 (2003): 20–21.

"The Colouration of the Extinct Cape Colony Quagga." *African Wildlife* 37, no. 4 (1983): 136–39.

"Correspondence: Does the Taxonomy of the Quagga Really Need to Be Reconsidered?" *South African Journal of Science* 93 (1997): 67–68.

"Das Quagga Und Seine Ruckzucht." *Natur und Museum: Bericht der Senckenbergischen Naturforschenden Gesellschaft* 127, no. 2 (1992): 37–45.

"How James Drury Cast the Bushmen Displayed at the South African Museum." *South African Journal of Science* 60, no. 8 (1964): 242–44.

"Quagga Experimental Breeding Project." In *Spirit of Enterprise: The 1990 Rolex Awards*. Edited by D. W. Reed, 434–36. Bern, Switerland: Buri International, 1990.

"Quagga Project Management/Action Plan." 2005. https://studylib.net/doc/ 7624988/–the-quagga-project.

"Quaggas in Museums all over the World." n.d. https://quaggaproject.org/ quaggas-in-museums-worldwide/.

"Revised List of the Preserved Material of the Extinct Cape Colony Quagga, *Equus quagga quagga* (Gmelin)." *Annals of the South African Museum* 65, no. 2 (1974): 41–87.

Rough Road Towards Re-Breeding the Quagga: How It Came About. Cape Town: Iziko South African Museum Library, 1999.

Ridgeway, W. "Contributions to the Study of the Equidae. On Hitherto Unrecorded Specimens of Equus quagga." *Proceedings of the Zoological Society of London* 38 (1909): 563–86.

Ritvo, H. *Noble Cows and Hybrid Zebras: Essays on Animals and History*. Charlottesville: University of Virginia Press, 2010.

The Platypus and the Mermaid and Other Figments of the Classifying Imagination. Cambridge, MA: Harvard University Press, 1997.

"Q Is for Quagga." In *Animalia: An Anti-Imperial Bestiary for Our Times*. Edited by A. M. Burton, 145–62. Durham, NC: Duke University Press, 2020.

Roberts, A. "Not Likely to be Quaggas. Mr. Austin Roberts and 'Discovery'." *The Star* (Johannesburg). June 20, 1932, 7.

The Mammals of South Africa. South Africa: Mammals of South Africa Book Fund, 1951.

Rookmaaker, L. C. *The Zoological Exploration of Southern Africa, 1659–1790*. Rotterdam: A. A. Balkema, 1989.

Rose, R. W. *Bushman, Whale and Dinosaur: James Drury's Forty Years at the South African Museum*. Cape Town: H. Timmins, 1961.

A Concise History of South Africa. Cambridge: Cambridge University Press, 1999.

Rowley, J. *The Art of Taxidermy*. New York: D. Appleton & Co., 1898.

Rubenstein, D., B. Low Mackey, Z. D. Davidson, F. Kebede, and S. R. B. King. "Equus grevyi." *The IUCN Red List of Threatened Species*, 2016. www .iucnredlist.org/species/7950/89 624491, accessed November 27, 2020.

Rubenstein, D. I., S. R. Sundaresan, I. R. Fischhoff, C. Tantipathananandh, and T. Y. Berger-Wolf. "Similar but Different: Dynamic Social Network Analysis

Highlights Fundamental Differences between the Fission-Fusion Societies of Two Equid Species, the Onager and Grevy's Zebra." *PLoS ONE* 10, no. 10 (2015), e0138645.

Ruxton, G. D. "The Possible Fitness Benefits of Striped Coat Coloration for Zebra." *Mammal Review* 32, no. 4 (2002): 237–44.

Sachs, R. "Live Weights and Body Measurements of Serengeti Game Animals." *African Journal of Ecology* 5, no. 1 (1967): 24–36.

Sandler, R. "The Ethics of Reviving Long Extinct Species." *Conservation Biology* 28, no. 2 (2014): 354–60.

Schlawe, L. *Über Die Ausgerotteten Steppenzebras Von Sudafrika Quagga Und Dauw, Equus quagga quagga.* Köln: Zoologischen Garten, 2010.

Schouman, A. *Een Quagga (Equus quagga).* Haarlem: Teylor's Museum, 1780. www.teylersmuseum.nl/nl.

Sclater, P. L. *Guide to the Gardens of the Zoological Society of London.* London: Bradbury, Agnew & Co., 1875.

Scott-Samuel, N. E., R. Baddeley, C. E. Palmer, I. C. Cuthill, and D. C. Burr. "Dazzle Camouflage Affects Speed Perception." *PLoS ONE* 6, no. 6 (2011): e20233.

Seddon, P. J., and D. P. Armstrong. "The Role of Translocation in Rewilding." In *Rewilding.* Edited by N. Pettorelli, S. M. Durant, and J. T. du Toit, 303–24. Cambridge: Cambridge University Press, 2019.

Seddon, P. J., A. Moehrenschlager, and H. Akcakaya. "IUCN Guiding Principles on Creating Proxies of Extinct Species for Conservation Benefit." *International Union for the Conservation of Nature*, 2016. https://library.wcs.org/doi/ctl/view/mid/33065/pubid/PUB19074.aspx.

Selous, F. C. "Burchell's Zebra." In *Great and Small Game of Africa.* Edited by H. A. Bryden, 79–84. London: R. Ward, 1899.

Selous, F. C., and T. Roosevelt. *African Nature Notes and Reminiscences . . . With a 'Foreword' by President Roosevelt and Illustrations by E. Caldwell.* London: Macmillan & Co., 1908.

Seuss, Dr. *If I Ran the Zoo.* New York: Random House Children's Books, 2014.

Siegfried, W. R. "Human Impacts." In *The Karoo: Ecological Patterns and Processes.* Edited by W. R. J. Dean and S. J. Milton, 239–41. Cambridge: Cambridge University Press, 1999.

Silverberg, R. *Born with the Dead: Three Novellas.* London: Gollancz, 1975.

Silverston, A. "Khumba." In *Wikipedia.* https://en.wikipedia.org/wiki/Khumba. Last accessed November 21, 2021.

Simmonds, P. S. *Animal Products: Their Preparation, Commercial Uses and Value.* London: Chapman & Hall, 1877.

Simmons, L. W., and M. Lovegrove. "Nongenetic Paternal Effects Via Seminal Fluid." *Evolution Letters* 3, no. 4 (2019): 403–11.

Simons, J. *The Tiger That Swallowed the Boy: Exotic Animals in Victorian England.* Faringdon, Oxfordshire: Libri Publishing, 2012.

Skead, C. J. *Historical Mammal Incidence in the Cape Province.* 2 vols. Cape Town: Department of Nature and Environmental Conservation of the Provincial Administration of the Cape of Good Hope, 1980–87.

Historical Mammal Incidence in the Cape Province, Volume 2: The Eastern Half of the Cape Province, Including the Ciskei, Transkei and East Griqualand. Cape Town:

The Chief Directorate, Nature and Environmental Conservation of the Provincial Administration of the Cape of Good Hope, 1987.

Skinner, J. D. "Further Light on the Speciation in the Quagga." *South African Journal of Science* 92 (1996): 301–2.

"Nowhere to Run: Mammals in Trouble." *African Wildlife* 59, no. 4 (2005): 16–21.

Skinner, J. D., and C. T. Chimimba. *The Mammals of the Southern African Subregion.* Cambridge: Cambridge University Press, 2005.

Smith, C. H. *The Natural History of Horses: The Equidae or Genus Equus of Authors. Naturalist's Library, Mammalia,* Vol. 12, London: W. H. Lizars, 1841.

Smith, R. K., E. Ryan, E. Morley, and R. A. Hill. "Resolving Management Conflicts: Could Agricultural Land Provide the Answer for an Endangered Species in a Habitat Classified as a World Heritage Site?" *Environmental Conservation* 38, no. 3 (2011): 325–33.

Smuts, G. L. "Pre- and Postnatal Growth Phenomena of Burchell's Zebra Equus burchelli antiquorum." *Koedoe: African Protected Area Conservation and Science* 18, no. 1 (1975): 69–102.

"Reproduction in the Zebra Mare Equus burchelli antiquorum from the Kruger National Park." *Koedoe: African Protected Area Conservation and Science* 19, no. 1 (1976): 89–132.

Somerville, W., E. Bradlow, and F. R. Bradlow. *William Somerville's Narrative of His Journeys to the Eastern Cape Frontier and to Lattakoe, 1799–1802.* Cape Town: Van Riebeeck Society, 1979.

South African National Parks. "Addo Elephant National Park." *South African National Parks,* n.d. www.sanparks.org/parks/addo/conservation/ff/mammals .php. "Karoo National Park."

Sparrman, A. *A Voyage to the Cape of Good Hope, Towards the Antarctic Polar Circle, and Round the World, but Chiefly into the Country of the Hottentots and Caffres, from the Year 1772 to 1776.* 2 vols. London: G. G. J. and J. Robinson, 1785–86.

The Spectator. "African Skins." *New York Times.* June 19, 1898, 17.

"Man the Destroyer." *New York Times.* January 9, 1887, 10.

Spreen, R. *Monument voor de Quagga: Schlemiel van de Uitgestorven Dieren.* Amsterdam: Fusilli, 2016.

Stalmans, M. E., T. J. Massad, M. J. S. Peel, C. E. Tarnita, and R. M. Pringle. "War-Induced Collapse and Asymmetric Recovery of Large-Mammal Populations in Gorongosa National Park, Mozambique." *PLoS ONE* 14, no. 3 (2019), e0212864.

Star, S. L. and J. R. Griesemer. "Institutional Ecology, 'Translations' and Boundary Objects: Amateurs and Professionals in Berkeley's Museum of Vertebrate Zoology, 1907–39." *Social Studies of Science* 19, no. 3 (1989): 387–420.

Stears, K., A. Shrader, and G. Castley. "Equus quagga – Plains Zebra." In *The Red List of Mammals of South Africa, Swaziland and Lesotho.* Edited by M. F. Child, L. Roxburgh, E. Do Linh San, D. Raimondo, and H. T. Davies-Mostert, 1–8. Midrand: National Biodiversity Institute and Endangered Wildlife Trust, 2016.

Steiner, C. C., and O. A. Ryder. "Molecular Phylogeny and Evolution of the Perissodactyla." *Zoological Journal of the Linnean Society* 163, no. 4 (2011): 1289–303.

Stewart, R., and B. Warner. "William John Burchell: The Multi-Skilled Polymath." *South African Journal of Science* 108, no. 11–12 (2012): 52–61.

Stokstad, E., R. P. M. A. Crooijmans, M. Upadhyay, and J. A. M. van Arendonk. "Bringing Back the Aurochs: By Conjuring the Extinct Ancestor of Modern Cattle, Breeders Are Making Europe Just a Little Wilder (Interview with Richard Crooijmans, Maulik Upadhyay and Johan Van Arendonk)." *Science* 350, no. 6265 (2015): 1144–47.

Storey, W. K. *Guns, Race, and Power in Colonial South Africa.* Cambridge: Cambridge University Press, 2008.

Streak, D. "Quagga Project Forced to Farm out Animals." *Sunday Times* (South Africa). September 26, 1993.

Strindberg, A. *The Father: A Tragedy.* The Project Gutenberg EBook of Plays. 1887. www.gutenberg.org/files/8499/8499-h/8499-h.htm#link2H_4_0007.

Sutton, T. *The Daniells. Artists and Travellers.* London: Bodley Head, 1954.

Swart, S. S. "Frankenzebra: Dangerous Knowledge and the Narrative Construction of Monsters." *Journal of Literary Studies* 30, no. 4 (2014): 45–70.

"Resurrection Conservation: The Return of the Extinct?" In *Nature Conservation in Southern Africa: Morality and Marginality: Towards Sentient Conservation?* Edited by J. B. Gewald, M. Spierenburg, and H. Wels, 130–64. Leiden: Brill, 2019.

Riding High: Horses, Humans and History in South Africa. Johannesburg: Wits University Press, 2010.

"Riding High: Horses, Power and Settler Society, c.1654–1840." *Kronos* 33, no. 29 (2003): 47–63.

"Zombie Zoology. History and Reanimating Extinct Animals." In *The Historical Animal.* Edited by S. Nance, 54–71. Syracuse, NY: Syracuse University Press, 2015.

Tegetmeier, W. B., J. C. Ewart, D. Barnaby, and S. Bartlett. *Letters to Mr Tegetmeier: From J. Cossar Ewart and Others to the Editor of the Field at the Turn of the Nineteenth Century.* Timperley, UK: ZSGM Publications, 2004.

Tegetmeier, W. B., and C. L. Sutherland. *Horses, Asses, Zebras, Mules and Mule Breeding.* London: H. Cox, 1895.

Thackeray, J. F. "Morphometric, Palaeoecological and Taxonomic Considerations of Southern African Zebras: Attempts to Distinguish the Quagga." *South African Journal of Science* 93, no. 2 (1997): 89–93.

"Zebras from Wonderwerk Cave, Northern Cape Province, South Africa: Attempts to Distinguish Equus burchelli and E. quagga." *South African Journal of Science* 84, no. 2 (1988): 99–101.

Thamm, E. "Rebirth of the Quagga." *African Wildlife* 5, no. 3 (1951): 209.

Thom, H. B. *Journal of Jan Van Riebeeck: 1659–1662.* 3 vols. Cape Town: The Van Riebeeck Society, 1952–58.

Townsend, J. R., and B. Pfeffer. *Modern Poetry: A Selection.* Philadelphia: Lippincott, 1974.

Trouessart, E. L. "Le Couagga et le Zèbre de Burchell de la Collection du Muséum." *Bulletin du Muséum National d'Histoire Naturelle* 7 (1906): 449–52.

Turnbull, M. "Back from the Dead." *Africa Environment & Wildlife* 9, no. 3 (2001): 30–37.

Turner Corporation. "Bontebok Ridge Reserve." 2020. https://turnercorporation .wixsite.com/turnercorp.

Van Bruggen, A. C. "The Last Quagga." *African Wildlife* 13, no. 4 (1959): 279–80.

Van der Merwe, P. J. *Migrant Farmer in the History of the Cape Colony, 1657–1842.* Translated by R. B. Beck. Athens: Ohio University Press, 1995.

Van Rensburg, E. "City Taxidermist Restores Quagga Exhibit in Germany." Cape Times. May 13, 1999.

Van Sittert, L. "Bringing in the Wild: The Commodification of Wild Animals in the Cape Colony/Province c.1850–1950." *The Journal of African History* 46, no. 2 (2005): 269–91.

Van Wyk, B. E., P. A. Novellie, and C. M. van Wyk. "Flora of the Zuurberg National Park. 1. Characterization of Major Vegetation Units." *Bothalia* 18, no. 2 (1988): 211–20.

Vilstrup, J. T., A. Seguin-Orlando, M. Stiller, A. Ginolhac, M. Raghavan, S. C. A. Nielsen, . . . L. Orlando. "Mitochondrial Phylogenomics of Modern and Ancient Equids." *PLoS ONE* 8, no. 2 (2013): e55950.

Von Helversen, B., L. J. Schooler, and U. Czienskowski. "Are Stripes Beneficial? Dazzle Camouflage Influences Perceived Speed and Hit Rates." *PloS ONE* 8, no. 4 (2013): e61173.

Waage, J. K. "How the Zebra Got Its Stripes: Biting Flies as Selective Agents in the Evolution of Zebra Coloration." *Journal of the Entomological Society of Southern Africa* 44, no. 2 (1981): 351–58.

Wallace, A. R. *Natural Selection and Tropical Nature: Essays on Descriptive and Theoretical Biology.* London: Macmillan & Co., 1878.

Watkeys, M. K. "Soils of the Arid South-Western Zone of Africa." In *The Karoo: Ecological Patterns and Processes.* Edited by W. R. J. Dean and S. J. Milton, 17–26. Cambridge: Cambridge University Press, 1999.

Watson, A., and J. Clifford. "Mapping Supply Chains for Nineteenth Century Leather." NiCHE: Network in Canadian History & Environment, 2014. http://niche-canada.org/2014/08/11/mapping-19th-century-leathers-supply-chains/.

Wedderwill, "The Quagga Project." 2011. https://wedderwill.co.za/blog/the-quagga-project/

Weel, S., L. H. Watson, J. Weel, J. A. Venter, and B. Reeves. "Cape Mountain Zebra in the Baviaanskloof Nature Reserve, South Africa: Resource Use Reveals Limitations to Zebra Performance in a Dystrophic Mountainous Ecosystem." *AJE African Journal of Ecology* 53, no. 4 (2015): 428–38.

Weir, H. "The Quagga in the Zoological Society's Gardens, Regents Park." The Illustrated London News, November 6, 1858, 427.

Weismann, A. *The Germ-Plasm. A Theory of Heredity.* Translated by W. N. Parker and H. Rönnfeldt. New York: Charles Scribner's Sons, 1893.

Wende, H., L. Gallagher, A. O'Connell, and P. B. Msimango. *The Quagga's Secret.* Umbilo, Durban: Gecko Books, 1995.

Whiten, A. "Cultural Evolution in Animals." *Annual Review of Ecology, Evolution, and Systematics* 50 (2019): 27–48.

Wildswinkel Boland Veiling. YouTube, 2017. www.facebook.com/watch/live/?ref=watch_permalink&v=1149124828543279, last accessed December 10, 2021.

Williams, S. D. "*Equus grevyi* Grévy's Zebra." In *Mammals of Africa. Vol. 5, Carnivores, Pangolins, Equids and Rhinoceroses.* Edited by J. Kingdon and M. Hoffmann, 422–28. London: Bloomsbury Publishing, 2013.

Wilman, M. *The Rock-Engravings of Griqualand West & Bechuanaland South Africa.* Cape Town: A. A. Balkema, 1968.

Wilmut, I., A. E. Schnieke, J. McWhir, A. J. Kind, and K. H. S. Campbell. "Viable offspring derived from fetal and adult mammalian cells." *Nature* 385 (1997): 810–13.

Wilson, D. "Making the Nēnē Matter: Valuing Life in Postwar Conservation." *Environmental History* 25, no. 3 (2020): 492–514.

Witz, L. "The Making of an Animal Biography: Huberta's Journey into South African Natural History, 1928–1932." *Kronos: Journal of Cape History* 13, no. 30 (2004): 138–66.

Yeld, J. "Re-Breeding of Quagga Moves into a New Phase." *Cape Argus* (South Africa). July 27, 1993.

Zola, É. *Madeleine Férat: A Realistic Novel.* London: Vizetelly & Co., 1889.

Index